Intelligent Systems Reference Library

Volume 120

Series editors

Janusz Kacprzyk, Polish Academy of Sciences, Warsaw, Poland
e-mail: kacprzyk@ibspan.waw.pl

Lakhmi C. Jain, University of Canberra, Canberra, Australia;
Bournemouth University, UK;
KES International, UK
e-mail: jainlc2002@yahoo.co.uk; jainlakhmi@gmail.com
URL: http://www.kesinternational.org/organisation.php

About this Series

The aim of this series is to publish a Reference Library, including novel advances and developments in all aspects of Intelligent Systems in an easily accessible and well structured form. The series includes reference works, handbooks, compendia, textbooks, well-structured monographs, dictionaries, and encyclopedias. It contains well integrated knowledge and current information in the field of Intelligent Systems. The series covers the theory, applications, and design methods of Intelligent Systems. Virtually all disciplines such as engineering, computer science, avionics, business, e-commerce, environment, healthcare, physics and life science are included.

More information about this series at http://www.springer.com/series/8578

Giner Alor-Hernández · Rafael Valencia-García
Editors

Current Trends on Knowledge-Based Systems

 Springer

Editors
Giner Alor-Hernández
Division of Research and Postgraduate
  Studies
Instituto Tecnológico de Orizaba
Orizaba, Veracruz
Mexico

Rafael Valencia-García
Departamento de Informática y Sistemas,
  Facultad de Informática
Universidad de Murcia
Murcia
Spain

ISSN 1868-4394 ISSN 1868-4408 (electronic)
Intelligent Systems Reference Library
ISBN 978-3-319-84775-7 ISBN 978-3-319-51905-0 (eBook)
DOI 10.1007/978-3-319-51905-0

Printed on acid-free paper

This Springer imprint is published by Springer Nature
The registered company is Springer International Publishing AG
The registered company address is: Gewerbestrasse 11, 6330 Cham, Switzerland

Preface

Knowledge-based technologies are gaining momentum and they are currently achieving a certain degree of maturity. They are especially valuable in situations in which the amount of available information is prohibitive for the intuition of an unaided human decision maker and in which precision and optimality are of importance. Knowledge-based systems can aid human cognitive deficiencies by integrating various sources of information, providing intelligent access to relevant knowledge, and aiding the process of structuring decisions. They can also support choice among well-defined alternatives and build on formal approaches, such as the methods of engineering economics, operations research, statistics analysis, and decision theory. They can also employ artificial intelligence methods to address heuristically problems that are intractable by formal techniques. They provide a consistent and reliable basis to face the challenges for organization, manipulation and visualization of the data and knowledge, playing a crucial role as the technological basis of the development of a large number of Computational Intelligence Systems.

These technologies draw on standard and novel techniques from various disciplines within Computer Science, including Knowledge Engineering, Natural Language Processing, Decision Support Systems, Artificial Intelligence, Databases, Software Agents, etc. The methods and tools developed and integrated for this purpose are generic and have a very large application potential in a large amounts of fields like Information Retrieval, Semantic Searches, Information Integration, Information Interoperability, Bioinformatics, eHealth, eLearning, Software Engineering, eCommerce, eGovernment, Social Networks, eSupply Chain, etc. This book considers the following industrial sectors, but is not limited: Aerospace, Agriculture, Automotive, Banking, Business Services, Food Manufacturing, Mining and Mineral Extraction, National Government, Insurance, Energy Services and others.

The aim of this book is to disseminate current trends among innovative and high-quality research regarding the implementation of conceptual frameworks, strategies, techniques, methodologies, informatics platforms and models for developing advanced Knowledge-based methods and techniques and their application in different fields. The specific objectives can be summarized as:

- Create a collection of theoretical, real-world and original research works in the field of Knowledge Based Systems.
- Go beyond the state-of-the-art in the field of Knowledge-Based Systems.
- Publish successful applications and use cases of new approaches, applications, methods, techniques for developing advanced Knowledge-Based Systems and their application in different fields.
- Provide an appropriate dissemination venue from both academia and industrial communities.

This book contains one kind of contribution: regular research papers. These works have been edited according to the norms and guidelines of Springer Verlag Editorial. Several call for chapters were distributed among the main mailing lists of the field for researchers to submit their works to this issue. In the first deadline, we received a total of 25 expressions of interest in the form of abstracts. Due to the large amount of submissions, abstracts were subject to a screening process to ensure their clarity, authenticity, and relevancy to this book. Proposals came from several countries such as Brazil, Colombia, India, Greece, India, Ireland, the Republic of Korea, Malaysia, Malta, Mexico, New Zealand, Norway, Philippines, Poland, Romania, Serbia, Spain, Taiwan, Tunisia, Turkey, United Kingdom of Great Britain, Northern Ireland, and United States of America.

After the screening process, 15 proposals were invited to submit full versions. At least two reviewers were assigned to every work to proceed with the peer review process. 13 chapters were finally accepted for their publication after corrections requested by reviewers and editors were addressed.

The book content is structured in three parts: (1) Semantic Web applications, (2) Knowledge Acquisition & Representation, (3) Knowledge-based Decision Support Systems (Tools for Industrial Knowledge Management).

Semantic Web Applications: This part contains four chapters.

Chapter 1, entitled *im4Things: An Ontology-based Natural Language Interface for controlling devices in the Internet of Things*, proposes a natural language interface for the Internet of Things, which takes advantage of Semantic Web technologies to allow non-expert users to control their home environment through an instant messaging application in an easy and intuitive way. Several experiments were conducted with a group of end users aiming to evaluate the effectiveness of the approach proposed to control home appliances by means of natural language instructions. The evaluation results proved that without the need for technicalities, the user was able to control the home appliances in an efficient way.

Chapter 2, entitled *Knowledge-Based Leisure Time Recommendations in Social Networks*, presents a novel knowledge-based recommendation algorithm for leisure time information to be used in social networks, which enhances the state-of-the-art in this algorithm category by taking into account (a) qualitative aspects of the recommended places (restaurants, museums, tourist attractions etc.), such as price, service and atmosphere, (b) influencing factors between social network users, (c) the semantic and geographical distance between locations and (d) the semantic categorization of the places to be recommended. The combination of these features leads to more accurate and better user-targeted leisure time recommendations.

Chapter 3, entitled *An Ontology based System for Knowledge Profile Management: A Case Study in the Electric Sector*, presents an ontology to help manage knowledge profiles in organizations. The ontology was implemented by using information obtained from a real case, where the roles and knowledge required for the people in charge of the processes of an electricity generation enterprise were analyzed.

Chapter 4, entitled *Sentiment Analysis based on Psychological and Linguistic Features for Spanish language*, presents an extensive experiments to evaluate the effectiveness of the psychological and linguistic features for sentiment classification. To this purpose, four psycholinguistic dimensions obtained from LIWC were used, and one stylometric dimension obtained from WordSmith, for the subsequent training of the SVM, Naïve Bayes, and J48 algorithms. A corpus of tourist reviews from the travel website TripAdvisor was created. The findings reveal that the stylometric dimension is quite feasible for sentiment classification. Finally, with regard to the classifiers, SVM provides better results than Naïve Bayes and J48 with an F-measure rate of 90.8%.

Knowledge Acquisition and Representation: This part contains four chapters.

Chapter 5, entitled *Knowledge-based System in an Affective and Intelligent Tutoring System*, proposes an affective and intelligent tutoring system called Fermat that integrates emotion or affective states with an Intelligent Learning Environment. The system applies Knowledge Space Theory to implement the knowledge representation in the domain and student modules and Fuzzy Logic to implement a new knowledge tracing algorithm, which is used to track student's pedagogical and affective states. The Intelligent Learning Environment was implemented with two main components: an affective and intelligent tutoring system for elementary mathematics and an educational social network. The tutoring system generates math exercises by using a fuzzy system that is fed with cognitive and effective values.

Chapter 6, entitled *A software strategy for knowledge transfer in a pharmaceutical distribution company*, presents an approach to solve knowledge transfer problems faced by a family owned pharmaceutical distribution company. The main objective is to improve knowledge transfer efficiency, recover outdated knowledge and improve the company's operation.

Chapter 7, entitled *GEODIM: A semantic model-based system for 3D recognition of industrial scenes*, presents GEODIM a semantic model-based system for recognition of 3D scenes of indoor spaces in factories. The system relies on the two technologies to describe industrial digital scenes with logical, physical, and semantic information. GEODIM extends the functionality of traditional object recognition algorithms by incorporating semantics in order to identify and characterize recognized geometric primitives along with rules for the composition of real objects. The research also describes a real case where GEODIM processes were applied and presents its qualitative evaluation.

Chapter 8, entitled *Beyond Interoperability in Critical Systems Engineering,* in this chapter a conceptual layer of interoperability is outlined describing what kind of features a powerful new interoperability technology should support in order to fuel desired changes in engineering and production paradigms.

Knowledge-Based Decision Support Systems: This part contains five chapters.

Chapter 9, entitled *Knowledge-based Decision Support Systems for Personalized u-lifecare Big Data Services,* proposes an architecture and case study of a Knowledge-based Big data acquisition, storage and processing platform for personalized u-lifecare services including data analytics and reasoning and inferencing services. Provides high performance computing for intensive data processing in cost effective manner. The main objective of the platform is to permit a systematic data management and effective utilization of the users' generated data to help users to visualize the personal behaviour patterns and to facilitate u-lifecare services to manage their daily routines.

Chapter 10, entitled *Decision support system for operational risk management in supply chain with 3PL providers,* presents a multicriteria decision support system for effective management of the operational risks present in a supply chain that includes 3PL providers, specifically in ground transportation of goods. The model is supported by Fuzzy QFD for the prioritization of risks in terms of their impact on the performance indicators that are considered relevant by the actors in the supply chain. Findings indicate that the proposed model allows prioritizing the risks according with the most important indicators.

Chapter 11, entitled *Expert System Development for the Assessment of Ergonomic Compatibility: Selection of Advanced Manufacturing Technology,* proposes the development of an expert system for ergonomic compatibility assessment on the selection of Advanced Manufacturing Technology (AMT). The research proposes a novel axiomatic design methodology under fuzzy environment including two stages: the generation of fuzzy If-Then rules using Mamdani's fuzzy inference system and the development of the system by mean of experts' opinions. A numerical example is presented for the selection of three CNC milling machines using the Weighted Ergonomic Incompatibility Content (WEIC).

Chapter 12, entitled *Developing Geo-recommender systems for Industry,* presents an integration architecture for developing a geo-recommender system. The architecture is composed of different layers, where the functionalities and interrelations of the layer components are distributed in order to ensure maintenance and scalability. A web-based system called GEOREMSYS was developed in order to recommend and to select Points Of Sale (POS).

Chapter 13, entitled *Evaluation of Denoising Methods in the Spatial Domain for Medical Ultrasound Imaging Applications,* presents the evaluation of denoising techniques, designed specifically for multiplicative noise models, applied in the spatial domain. The evaluation is analyzed and compared by using a synthetic image, a phantom image and real images. The aim of this study is to compare denoising methods when no transformation of the image is carried out.

Once a brief summary of chapters has been provided, we would also like to express our gratitude to the reviewers who kindly accepted to contribute in the evaluation of chapters at all stages of the editing process.

Orizaba, Mexico   Giner Alor-Hernández
Murcia, Spain   Rafael Valencia-García

Acknowledgements

Guest editors will always be grateful for the talented technical reviewers who helped review and improve this book. The knowledge and enthusiasm they brought to the project was simply amazing.

Thus, we would like to thank:

To all our colleagues and friends from the Instituto Tecnológico de Orizaba and Universidad de Murcia for all their support.

We equally and especially wish to thank Springer Verlag and associate editors of Intelligent Systems Reference Library book series, for grating us the opportunity to edit this book and providing valuable comments to improve the selection of research works.

Guest editors are grateful to the National Technological of Mexico for supporting this work. This book was also sponsored by the National Council of Science and Technology (CONACYT) as part of the project named Thematic Network in Industrial Process Optimization, as well as by the Public Education Secretary (SEP) through PRODEP. Finally, this book has been partially supported by the Spanish Ministry of Economy and Competitiveness and the European Commission (FEDER/ERDF) through project KBS4FIA (TIN2016-76323-R).

Contents

Contributors

Giner Alor-Hernández Division of Research and Postgraduate Studies, Technological Institute of Orizaba, Orizaba, Veracruz, Mexico

Liliana Avelar-Sosa Autonomous University of Ciudad Juarez, Ciudad Juarez, Chihuahua, Mexico

Thar Baker Department of Computer Science, Faculty of Engineering and Technology, Liverpool John Moores University, Liverpool, UK

Mario Barcelo-Valenzuela Universidad de Sonora, Hermosillo, Sonora, Mexico

Ramón Zatarain Cabada Departamento de Posgrado, Instituto Tecnológico de Culiacán, Culiacán, Sinaloa, Mexico

Patricia Shihemy Carrillo-Villafaña Universidad de Sonora, Hermosillo, Sonora, Mexico

Ángel García Crespo Computer Science Department, Universidad Carlos III de Madrid, Leganés, Madrid, Spain

Jose Luis López Cuadrado Computer Science Department, Universidad Carlos III de Madrid, Leganés, Madrid, Spain

María Lucía Barrón Estrada Departamento de Posgrado, Instituto Tecnológico de Culiacán, Culiacán, Sinaloa, Mexico

Muhammad Fahim Department of Computer Engineering, Faculty of Engineering and Natural Sciences, Istanbul Sabahattin Zaim University, Istanbul, Turkey

Jorge Luis García-Alcaraz Departamento de Ingeniería Industrial y Manufactura, Instituto de Ingeniería y Tecnología, Universidad Autónoma de Ciudad Juárez, Ciudad Juárez, México

Vicente García Jiménez Departamento de Ingeniería Eléctrica y Computación, Universidad Autónoma de Ciudad Juárez, Ciudad Juárez, Chihuahua, México

Panagiotis Georgiadis Department of Informatics and Telecommunications, University of Athens, Athens, Greece

Cynthya García de Jesús Computer Science Department, Universidad Carlos III de Madrid, Leganés, Madrid, Spain

Jorge Limón-Romero Universidad Autónoma de Baja California, Ensenada, Baja California, México

Aleksander Lodwich Department of Computer Science, Carlos III University of Madrid, Madrid, Spain

Aide Maldonado-Macías Universidad Autónoma de Ciudad Juárez, Ciudad Juárez, Chihuahua, México

Diego Fernando Manotas Duque Grupo de Investigación en Logística y Producción, Escuela de Ingeniería Industrial, Universidad del Valle, Cali, Colombia

Dionisis Margaris Department of Informatics and Telecommunications, University of Athens, Athens, Greece

Jose María Alvarez-Rodríguez Department of Computer Science, Carlos III University of Madrid, Madrid, Spain

José Ángel Noguera-Arnaldos Proyectos Y Soluciones Tecnológicas Avanzadas, SLP (Proasistech) Edificio CEEIM, Murcia, Spain

Humberto de Jesús Ochoa Domínguez Departamento de Ingeniería Eléctrica y Computación, Universidad Autónoma de Ciudad Juárez, Ciudad Juárez, Chihuahua, México

José Luis Ochoa Department of Industrial Engineering, Universidad de Sonora, Hermosillo, Mexico

Juan Carlos Osorio Gómez Grupo de Investigación en Logística y Producción, Escuela de Ingeniería Industrial, Universidad del Valle, Cali, Colombia

Mario Andrés Paredes-Valverde Department of Informatics and Systems, Universidad de Murcia, Murcia, Spain

Yuliana Perez-Gallardo Computer Science Department, Universidad Carlos III de Madrid, Leganés, Madrid, Spain

Alonso Perez-Soltero Universidad de Sonora, Hermosillo, Sonora, Mexico

Susana Itzel Pérez-Rodríguez Universidad Popular Autónoma Del Estado de Puebla, Puebla, Mexico

Yasmín Hernández Pérez Instituto Nacional de Electricidad y Energías Limpias, Tecnologías de la Información, Cuernavaca, Mexico

Arturo Realyvásquez Autonomous University of Ciudad Juarez, Ciudad Juarez, Chihuahua, Mexico

Leonardo Rivera Grupo de Investigación en Logística y Producción, Escuela de Ingeniería Industrial, Universidad del Valle, Cali, Colombia

Oscar M. Rodríguez-Elias División de Estudios de Posgrado e Investigación, Tecnológico Nacional de México—InstitutoTecnológico de Hermosillo, Hermosillo, Sonora, Mexico

Miguel Ángel Rodríguez-García Computational Bioscience Research Center, King Abdullah University of Science and Technology, Thuwal, Kingdom of Saudi Arabia

Lisbeth Rodríguez-Mazahua Division of Research and Postgraduate Studies, Instituto Tecnológico de Orizaba, Orizaba, Veracruz, Mexico

Cesar E. Rose-Gómez División de Estudios de Posgrado e Investigación, Tecnológico Nacional de México—InstitutoTecnológico de Hermosillo, Hermosillo, Sonora, Mexico

María Pilar Salas-Zárate Department of Informatics and Systems, Universidad de Murcia, Murcia, Spain

Gerardo Sanchez-Schmitz Universidad de Sonora, Hermosillo, Sonora, Mexico

José Luis Sánchez-Cervantes CONACYT-Instituto Tecnológico de Orizaba, Orizaba, Veracruz, Mexico

Cuauhtémoc Sánchez-Ramírez Division of Research and Postgraduate Studies, Instituto Tecnológico de Orizaba, Orizaba, Veracruz, Mexico

Rafael Valencia-García Department of Informatics and Systems, Universidad de Murcia, Murcia, Spain

Costas Vassilakis Department of Informatics and Telecommunications, University of the Peloponnese, Tripoli, Greece

Maria de Jesús Velázquez-Mendoza División de Estudios de Posgrado e Investigación, Tecnológico Nacional de México—InstitutoTecnológico de Hermosillo, Hermosillo, Sonora, Mexico

Edith Verdejo-Palacios Division of Research and Postgraduate Studies, Instituto Tecnológico de Orizaba, Orizaba, Veracruz, Mexico

List of Figures

List of Tables

Part I
Semantic Web Applications

Chapter 1
im4Things: An Ontology-Based Natural Language Interface for Controlling Devices in the Internet of Things

José Ángel Noguera-Arnaldos, Mario Andrés Paredes-Valverde, María Pilar Salas-Zárate, Miguel Ángel Rodríguez-García, Rafael Valencia-García and José Luis Ochoa

Abstract The Internet of Things (IoT) offers opportunities for new applications and services that enable users to access and control their working and home environment from local and remote locations, aiming to perform daily life activities in an easy way. However, the IoT also introduces new challenges, some of which arise from the large range of devices currently available and the heterogeneous interfaces provided for their control. The control and management of this variety of devices and interfaces represent a new challenge for non-expert users, instead of making their life easier. Based on this understanding, in this work we present a natural language interface for the IoT, which takes advantage of Semantic Web technologies to allow non-expert users to control their home environment through an instant messaging application in an easy and intuitive way. We conducted several experiments with a group of end users aiming to evaluate the effectiveness

J.Á. Noguera-Arnaldos
Proyectos Y Soluciones Tecnológicas Avanzadas, SLP (Proasistech) Edificio CEEIM,
Campus de Espinardo, 30100 Murcia, Spain
e-mail: jnoguera@proasistech.com

M.A. Paredes-Valverde (✉) · M.P. Salas-Zárate · R. Valencia-García
Department of Informatics and Systems, Universidad de Murcia, Murcia, Spain
e-mail: marioandres.paredes@um.es

M.P. Salas-Zárate
e-mail: mariapilar.salas@um.es

R. Valencia-García
e-mail: valencia@um.es

M.Á. Rodríguez-García
Computational Bioscience Research Center, King Abdullah University of Science
and Technology, 4700 KAUST, P.O. Box 2882, 23955-6900 Thuwal,
Kingdom of Saudi Arabia
e-mail: miguel.rodriguezgarcia@kaust.edu.sa

J.L. Ochoa
Department of Industrial Engineering, Universidad de Sonora, Hermosillo, Mexico
e-mail: joseluis.ochoa@industrial.uson.mx

© Springer International Publishing AG 2017
G. Alor-Hernández and R. Valencia-García (eds.), *Current Trends on Knowledge-Based Systems*, Intelligent Systems Reference Library 120,
DOI 10.1007/978-3-319-51905-0_1

3

of our approach to control home appliances by means of natural language instructions. The evaluation results proved that without the need for technicalities, the user was able to control the home appliances in an efficient way.

Keywords Natural language interface · Internet of things · Semantic Web

1.1   Introduction

The Internet of Things (IoT) refers to the pervasive presence around us of a variety of things or objects such as Radio-Frequency Identification (RFID), sensors, actuators, mobile phones, which, through unique addressing schemes, are able to interact with each other and cooperate with their neighboring smart components to reach common goals [1]. The IoT has been applied in several areas such as transportation systems [2], infrastructure construction, public security, environment protection, intelligent industry, urban management, among others [3]. One of the most attractive markets for the IoT application is the home automation, which refers to the application of computer and information technology for control of home appliances and domestic features [4].

Nowadays, the interaction with home appliances is mainly done through user interfaces based on mechanical buttons, knobs, and more recently, touch sense based buttons. These interfaces only allow users to turn on, turn off, or programming the corresponding device, thus limiting the functionalities that a device can provide, e.g., warnings, fault indication, the cooperation among them, and even, the actuation from remote locations. In this sense, the IoT offers the potential of endless opportunities for new applications and services in the home context that enable users to access and control the home environment, from local and remote locations, in order to carry out daily life activities in an easy way. All that previously mentioned improves the quality of life of the user while at the same time enabling energy efficiency. Despite these advantages, the large range of IoT devices and smart appliances often results to complex systems-of-systems interactions [5], i.e., the wide set of home appliances with diverse requirements and interaction patterns makes the smart home a challenging environment for the design and implementation of intuitive and easy-to-use applications for the end-users. Taking into account that the smart home control system plays a key role on the improvement of the system service quality and the interaction between smart home system and the users [6], it is imperative to provide users with a system that allows them to monitor and control the home environment, from local and remote locations in an easy and intuitive way. Regarding this fact, in this work we present an IoT-based system, known as im4Things, that enables users to control home appliances of different makes, models, and manufacturers, from local and remote locations by means of instant messages based on natural language. The im4Things system combines current technologies such as mobile applications, Natural Language Processing (NLP), Cloud computing, Semantic Web, and Raspberry Pi, an open-source platform used for building electronics projects.

Taking into account that a smart home control system should be affordable, scalable so that new devices can be easily integrated into the system, and it should be user-friendly [7], the im4Things system deals with three main issues. First, the need for an intuitive mechanism for home appliances control. In this sense, the im4Things system provides an instant messaging mobile application which allows users to control the home appliances through natural language (NL) messages. The NL paradigm is generally deemed to be the most intuitive from a usage point of view [8]. Also, the im4Things system implements a conversational agent in order to provide to users, in an enjoyable way, information concerning home appliances such as warnings, fault indication, as well as the different states that a home appliance can have during the execution of a user's instruction. Second, the representation and storing of the exchanged information in the context of the Internet of things. This fact is one of the most challenging issues on the IoT context [3]. In this sense, im4Things adopts semantic technologies, particularly ontologies, which represent appropriate modeling solutions for things description, reasoning over data generated by IoT, and communication infrastructure. An ontology is a formal and explicit specification of a shared conceptualization [9]. The ontologies have been successfully applied in domains such as cloud services [10], recommender systems [11], innovation management [12], and sentiment analysis [13], to mention but a few. Third, the need for smart things with proactive behavior, context awareness, and collaborative communications capabilities. In this regard, the im4Thing system provides a low-cost smart item, called im4Things bot, equipped with wireless communication, sensors, memory and elaboration capabilities. Thanks to these features, the im4Things bot not only receives data from home appliances and commands from application platform but also transmits data to the application platform, thus becoming in a generator and receiver of information.

The remainder of the paper is structured as follows. Section 1.2 presents a review of the literature about home automation and the IoT. The architecture design of the proposed approach, components and interrelationships are described in Sect. 1.3. Section 1.4 presents the evaluation results concerning the effectiveness of the present approach to control home appliances by means of written natural language instructions. Finally, conclusions and future work are presented.

1.2   Related Works

Recently, new approaches for home appliances control have been proposed, many of them targeted at the idea of providing speech-controlled home automation. The DIRHA (Distant-speech Interaction for Robust Home Applications) [14] is a system based on the detection and recognition of spoken commands preceded by a small key-phrase, which consist of one up to three words. The current prototype works with a single grammar that recognizes one key-phrase among possible garbage segments. DIRHA recognizes a set of 99 commands of various lengths and, in this regard, a finite-state-grammar was built. In [15] the authors presented G.H.O.

S.T., a voice control system for smart buildings. This system was implemented with the KNX technology, a standardized network communications protocol for intelligent buildings. In G.H.O.S.T., the control of the building is performed from a PC via a KNX/IP router and a KNX communication Fieldbus. Also, this interaction is carried out through predefined voice commands, which are focused on the control of lighting, sun-blinds, and air conditioning. On the other hand, Mayordomo [16] is a multimodal (oral, written and a GUI interface) dialogue system focused on the interaction with an Ambient Intelligence environment implemented in a home. Mayordomo implements an ASR (Automatic Speech Recognition) engine as well as a frames-based [17] understanding component. In order to generate answers to the users, Mayordomo makes use of a set of patterns containing information such as home appliance, room, status, among others.

There are some approaches that allow home appliances controlling via the Internet, either e-mail or a Web application. In [4] the authors presented a Raspberry Pi-based home automation system. This system allows the home appliances control through e-mail whose subject represents the instruction to be executed by the Raspberry Pi. The subject must consist of the word ON or OFF followed by a number, which indicates the home appliance to turn on or turn off. Once the corresponding command is executed, the system notifies the user the status of the work done. On the other hand, in [6] the authors presented a smart control system based on the Internet of Things. The system is based on the CoAP [18] protocol, which enables very simple electronics devices the communication over the Internet. This system enables the user to query and control several devices through the Web. For this, the system provides a Web interface with a set of predefined functions, each of which is triggered through a button element.

On the other hand, there are systems focused on home appliances control by means of mobile applications. For instance, in [19] the authors presented a BLE-based (Bluetooth Low Energy) [20] appliance control system. This system implements a "Middle-Device" for the communication between the user, through a mobile application, and home appliances using Arduino, an open-source platform used for building electronics projects. The home appliances control is performed through the selection of predefined functions available on the mobile application. Another example of this kind of systems is presented in [21]. In this work, the authors implemented an Arduino Ethernet based micro-web server for accessing and controlling appliances remotely from an Android application. The mobile application provides a GUI with a set of button elements each of which triggers a specific home appliance action, such as turn on the emergency light, to active the house alarm, and to lock the main door, among others. On the other hand, in [22] the authors presented a smart switch control system. This system provides a GUI implemented using Web technologies but packed as native applications. Also, it implements a SOC (System on a Chip) power metering chip through which the switching operation over saving lamps, air conditioners, refrigerators, and microwaves is performed.

Despite that the above-presented works have made a significant contribution to the smart home field, they present certain disadvantages that are addressed by our approach. First, works such as [19] implements Bluetooth technology. Despite Bluetooth capabilities are good and most of current smartphones and tablets have integrated this technology, wherewith the cost of the system reduces significantly, it limits the home appliance control within the Bluetooth range of the environment. In the same context, some works such as Mayordomo, DIRHA, and G.H.O.S.T. represent smart home systems which are intended to provide the control of the home environment just from inside the home. On the basis of this scenarios, the present work provides users the advantage of controlling the home environment not just from inside the home but also from remote locations thanks to a Cloud service implemented by the im4Things system. This feature is interesting for busy families or individuals for whom to take the control of the home environment from remote locations can represent time-saving, thus improving their quality of life. Second, some approaches such as Mayordomo, [6, 19, 21, 22] provide a GUI which must be modified when a new device is added to the smart home system. This fact can confuse to the users due to the continuous changes on the user interface. Also, these changes increase the development effort. In order to address the issue above, the im4Things system provides a mobile application that points to both configure a new device via wireless connection (Bluetooth) and, once the new device is configured, establish communication with the device by means of instant messages based on natural language. In this sense, the user does not perceive any changes on the user interface, due to the new device is added as a new contact, as it occurs in current instant messaging applications such as WhatsApp and Telegram. Third, the user interaction in systems such as DIRHA, G.H.O.S.T., and [4] is performed through the recognition of small key phrases, predefined voice commands, and through an e-mail message whose subject contains the keyword ON/OFF followed by a number, respectively. This fact can be tedious and complicated for end-users due to the need to learn a set of phrases or keywords. In this sense, the present approach provides an instant messaging application through which the users can interact with the home environment through natural language, thus avoiding the need to learn new commands. Also, as was previously mentioned, the natural language paradigm is generally deemed to be the most intuitive from a usage point of view. Besides, the im4Things system provides a conversational agent aiming to interact with the user through natural language, thus enhancing the user experience. Finally, it should be mentioned that the implemented conversational agent exploits the use of ontologies in order to address one of the main limits of this technology, the rigidness of the dialogue mechanism. All above-mentioned features are described in detail in next section.

1.3   im4Things System

According to [23], a generic NLI (Natural Language Interface) architecture requires four main components: (1) input devices that include user interfaces as input data mechanism, (2) a natural language understanding component that analyses the user input and, optionally interacts with domain-specific information, (3) an interaction manager component that maintains information about ongoing interaction, and (4) a natural language generation component that produces the appropriate responses. On the basis of this understanding, in this work we propose an IoT system, known as im4Things, which addresses all above-mentioned features through the implementation of three main components: a mobile application, a Cloud-based service, and a low-cost smart electronic device. The mobile application serves as the input data mechanism through which users interact with the system in natural language aiming to control the home environment. The Cloud-based service serves two functions: managing the interaction between the users and the home environment, and the natural language understanding. The first function is performed by means of the establishment of a bi-directional communication between the entities involved. This communication is based on the Extensible Messaging and Presence Protocol (XMPP) [24], an open and extensible XML-based protocol. On the other hand, the Natural Language understanding function aims to process the users' messages in order to detect the home appliance and the action it must execute. This processing is based on an ontology that describes the home appliances domain. Finally, the smart electronic device, known as im4Things bot, serves not only as a bridge between the application platform and the home environment but also as a natural language generator that provides information about the status of a home appliances during the execution of a task as well as the home environment status. Concerning this natural language generation process, the im4Things bot implements an AIML-based (Artificial Intelligence Markup Language) [25] dialog system.

In a nutshell, the im4Things system works as follows. First, the user adds and configures a new home appliance through the mobile application. Actually, the configuration process occurs between the mobile application and the im4Things bot that control the corresponding home appliance. This configuration is performed by means of Bluetooth connection. Once the new device is configured, the user, which automatically becomes the administrator of the device, is able to interact with it through natural language messages. Also, the administrator is able to provide access to the device to other users who, in addition to controlling the home appliance, also will be informed about any change on the home appliance's status. Then, when a user needs performing a housework by means of a specific device, the user sends a message to the corresponding im4Things bot. Before the message is received by the target device, it is processed by the im4Things Cloud service through the understanding module. This process aims to obtain all semantic information from the words contained in the message and determine the command that must be executed by the target device. As previously mentioned, this natural language processing is based on a home appliances ontology, which describes, among other things, the states, sensors, actions, alerts and services that a home appliance provides. When the im4Things bot receives the message, it interacts with the home appliance in order to execute the corresponding action. Finally, once the home appliance executes the command, the im4Things bot notifies the user, on natural language, about the result (success or failure) of the command execution.

Besides the above-mentioned process, the im4Things bot is also able to notify the users about the home appliance's state changes. Therefore, once a home appliance's alert is triggered, the im4Things bot sends to the im4Things Cloud service a natural language expression concerning the situation. Then, the im4Things Cloud service notifies the users subscribed to the home appliance. Finally, it is worth mentioning that im4Things system allows users to control home appliances from the local network and remote locations.

Figure 1.1 shows the architecture proposed in this work which, as previously mentioned, is composed of three main modules, (1) the im4Things App, (2) the im4Things Cloud Service, and (3) the im4Things Bot. These modules are described in detail in the next sections.

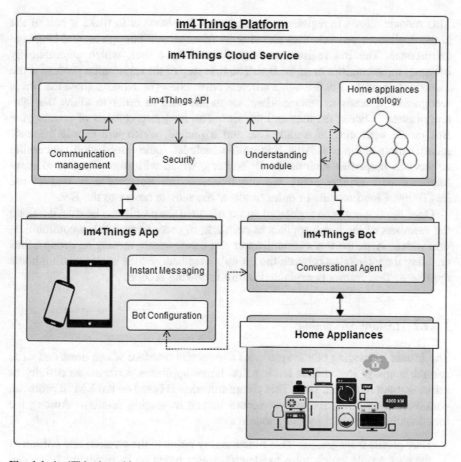

Fig. 1.1 im4Thing's architecture

1.3.1   im4Things App

The im4Things App is a mobile application through which users can establish a communication with different entities, either humans or Bots. The functionalities provided by this application can be grouped into two main categories: (1) Bot Configuration, which are related to the Bot's registration and configuration, and (2) Instant Messaging, which concern to the sending and receiving of messages to and from humans and Bots. In the following sections, the functionalities groups above are described in detail.

1.3.1.1   Bot Configuration

This module allows to register and configure a Bot in order to make it part of the home appliances network, thus allowing its management through natural language instructions. The Bot registration is performed by a user, which automatically becomes the administrator of the Bot. The first step of the registration process is the identification of the Bot through a wireless connection (Bluetooth). Once the Bot is recognized, it sends its unique identifier to the user in order to allow the synchronization between the user and the Bot. The final step consists of creating the Bot profile consisting of a nickname and a picture, which will enable the easy identification of the Bot. Also, this profile includes some information concerning connection properties such as the IP address, which will allow the Bot to communicate with another entity via the Internet. The Bot profile is sent to the im4Things Cloud module in order to allow the remote access to the Bot.

Once the Bot has been registered and configured, its profile can be shared among the members of the home appliances network, in order to establish a communication with it. Also, the Bot's administrator can create instant messaging groups, thus enabling the interaction between the group's members and the corresponding home appliance. This feature is explained in the following section.

1.3.1.2   Instant Messaging

The Instant Messaging module provides a graphical interface where users can write natural language instructions aiming that home appliance performs an activity or informs about its current state. This communication is based on the XMPP protocol, which allows the management of secure instant messaging sessions. Among the functionalities provided by this module are:

- User profile management. This functionality refers to the creation and edition of the user profile which helps to identify a user based on its nickname or avatar.
- Instant messaging. It allows the communication among the entities through instant messages.

- Group chat. This feature allows the administrator to control several home appliances from a unique chat window. Also, by means of this functionality, the administrator can add people to the group, thus allowing them to control all home appliances members of the group.
- User's presence. It refers to the fact that a user has agreed that another entity is allowed to view its presence, thus allowing the communication establishment among them. This feature is especially important in order to allow that a Bot can be visible or hidden to a certain group of users.

1.3.2 im4Things Cloud Service

The im4Thing system has been implemented on a Cloud-based architecture in order to take advantage of Cloud computing technology such as low costs, immediate access to hardware resources, scalability, [26] and, especially, the ability to deliver the im4Things system as a service over the Internet, thus allowing users to control their home appliances from remote locations. The im4Things Cloud service is composed of three main modules: the communication management module, security module, and the understanding module. The interaction between the entities, whether user or im4Things bot, and the im4Things Cloud service is performed through a REST-based API. All the modules above are described in next sections.

1.3.2.1 im4Things API

The im4Things API is a REST-based application programming interface which describes the services provided by the im4Thing Cloud such as the entities' management, authentication, and instant messaging. Through this API, both im4Things App and im4Things bot interact with the im4Things Cloud service.

1.3.2.2 Communication Management

The main objective of this module is the communication management among the system's entities. This communication is based on MongooseIM, a customizable and scalable platform for instant messaging for social media, gaming and telecommunications. In this sense, the present module takes advantage of several MongooseIM features such as multi-user chat, Websockets and privacy.

In the im4Things system, the communication can be established in the following different ways: human to human, human to bot, and bot to human. Therefore, one of the most critical points of this module is ensuring the quality and continuity of the service through a correct management of the instant messaging traffic. In this sense, we established a priority queue for the instant messaging management based on the participating entities. The present module distinguishes three priority levels:

- Priority 0. This level refers to the human-to-human interaction. In this case, the message is transmitted as it is.
- Priority 1. This level corresponds to the human-to-bot interaction. In this level, the message is interpreted as an instruction to the bot, therefore, it is analyzed in order to detect the target bot as well as the specific command to be executed by the latter.
- Priority 2. The higher priority level corresponds to the bot-to-human interaction. In this case, the message is interpreted as a bot notification, therefore, it is sent to all users subscribed to the bot. The bot-to-human interaction has the higher priority because the bot can notify the users about warning or alerts that demand the instant or quick participation of the im4Things bot's administrator.

1.3.2.3 Security

The im4Things approach uses a token-based authentication method. In this context, a token contains the security identity of an entity which allows the latter to be identified by the server. Therefore, each request made by an entity must contain the token. Also, in order to ensure the data security, all messages are encrypted using SSL certificates such as 2048 bits and SHA-2.

1.3.2.4 Understanding Module

The understanding module is based on a previous work [26] of our research group, which aims to classify a natural language question. The above-mentioned work uses an ontological model for representing all semantic information obtained from the question and for inferring the answer type expected by the user. The present module, for its part, aims to classify the kind of instruction that a user has provided, i.e., this module determines if the instruction is a command that must be executed by the home appliance or just a query of the state of the home appliance. Moreover, this module aims to determine the home appliance that must execute the instruction provided. This is particularly important due to, as was previously mentioned, the instant messaging application allows the creation of chatting groups, where a user can control several devices through a single window. In the chatting groups, it is not necessary to specify the home appliance that must execute the action since the understanding module infers the home appliance based on the semantic information extracted from the message as well as the home appliances ontology.

Aiming to reach the above-mentioned objective, the understanding module performs the natural language processing of the instruction in order to obtain all semantic information from the words contained within it. The most relevant natural language processing techniques implemented by this module are the named entity recognition (NER), lemmatization, and synonym extension. The NER technique seeks to locate and classify words contained in the user's instruction into predefined categories such as home appliances, sensors, alerts, and states, among others. It is worth noting that this technique makes use of the home appliances ontology in order to recognize and classify these entities. With regard to the lemmatization and synonym extension techniques, they are used to deal with user's instructions whose elements are not the same contained in the knowledge base (home appliances ontology), e.g. a word could be declared in different verb time, in plural, or it could represent a synonym of some entity contained in the knowledge base. The implementation of these techniques allows reducing cases where the mapping between words contained in the instruction and elements of the knowledge base fails, thus increasing the effectiveness of the system.

Based on the collected information, the system is able to determine the target home appliance, the command it must execute, as well as if the instruction can be executed by the home appliance. On the one hand, the determination of the home appliance and the command becomes a determinant factor in the chatting groups with more than one bot. For instance, let us consider a case where a user is the administrator of two home appliances, a coffee maker and an irrigation system, and the user uses a chatting group to control both appliances. When the user writes the instruction "please make a coffee", the system recognizes, based on the home appliances ontology, a semantic relation between the word "coffee" and the coffee maker. Therefore, the system sends this instruction to the coffee maker instead of the irrigation system. On the other hand, as mentioned above, the system is able to identify if the instruction provided by the user can be executed by the home appliance. For instance, if a user asks the lighting control to make a coffee at the chat window exclusively focused on the control of this appliance, the system will send a message to the user indicating the error. It must be noted that the above-mentioned interaction is performed through natural language.

Home appliances ontology

The home appliances ontology provides a formal representation of the environment in which the im4Things system is placed, i.e. the home appliances domain. In this sense, the home appliances ontology establishes five main classes:

- State. This class represents all possible states that can have a home appliance.
- Sensor. It represents all sensor available in the home appliance.
- Action. This class represents all functionalities provided by the home appliance.
- Alert. It represents the alerts that a home appliance can trigger.
- Service. This class represents all services that a home appliance can perform.

Figure 1.2 shows an excerpt of the home appliances ontology describing the coffee maker context. This ontology excerpt shows the six states that a coffee maker can have. Also, the coffee maker sensors (WaterLevel, Temperature, CoffeeLevel) are shown, these sensors can trigger different alerts (NoCoffee, NoWater, Drink_Ready), for instance, when the water tank is empty, the coffee maker triggers

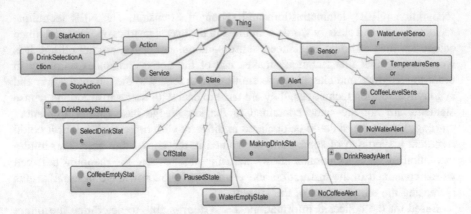

Fig. 1.2 An excerpt of the home appliances ontology describing the coffee maker's context

an alert asking the user refilling the tank. It has to be borne in mind that this communication is established through natural language messages. This process will be discussed next.

1.3.3 im4Things Bot

The im4Things Bot represents the interface between the home appliance and the im4Things top layers. The im4Things Bot has a specific hardware composed of sensors and a control and communication system that interacts with the software in order to execute commands and inform about the home appliance status in an efficient way. This bot is based on the Raspberry Pi hardware, which has been successfully applied in multiple control and voice recognition systems [27]. Also, the im4Thing bot embeds an AIML-based conversational agent which is responsible for simulate a human communication behavior in order to interact with the users through natural language. In the next section, this module is explained in detail.

1.3.3.1 Conversational Agent

The conversational agent is based on the dialog system technology. This technology was used to implement a system intended to converse with a human, with a coherent structure. In this sense, the AIML language was used to develop a natural language software agent. The AIML language is able to recognize predefined patterns in a text, and generate answers according to several conditions and states of the system. The AIML language is composed of regular expression patterns associated to a set of answers. These answers are provided to the user when the

```
<category>
<pattern>Are you a coffee maker? </pattern>
<template>
Yes, I am the most intelligent coffee maker in the world
<template>
</category>
<category>
<pattern>* is your name? </pattern>
<template>
My name is <bot name="name"/>, nice to meet you.
<template>
<category>
```

Fig. 1.3 Definition of two AIML patterns

corresponding pattern is identified. Figure 1.3 shows an excerpt of the definition of a conversational agent. This definition contains two patterns, the first one establishes that if the system identifies the phrase "Are you a coffee maker?", it will provide the answer defined by the "template" element. With regard to the second pattern, it is used when the system identifies the phrase "is your name?" preceded by several (*) words.

As described above, the AIML language is used to define patterns which produce an answer according to a simple correspondence of words. The rules (pattern matching rules) on which these conversational agents are based to carry out the dialogue are too restrictive and their language understanding capability is very limited [28]. On the basis of this understanding, we have implemented an extended version of the AIML-based structure above shown. The proposed structure takes into account the concepts contained in the home appliances ontology instead of a set of tokens (words). In this sense, the "pattern" element was extended aiming to represent the action, status, and the home appliance identified in the user's instruction. Also, the "action" element was added to the AIML structure; this element indicates the command that the home appliance must execute. Finally, the "template" element includes the answers that the system provides when the command is successfully executed or not. Figure 1.4 shows an excerpt of the Coffee maker's AIML-based patterns. The first pattern represents that the "DrinkSelection" action was found in the user's instruction and that the coffee maker's state is "Pause". Thus, when both conditions are met, the "MAKE_COFFEE" command will be sent to the home appliance. If the command is successfully executed, the system sends to the user the answer contained in the "success" element, otherwise, the system sends the answer contained in the "fail" element. The second pattern refers to a user's question about the state of the home appliance. Finally, it should be remembered that the system is also able to monitor the state of the home appliance and send an alert when there is a problem with it. In this sense, the third pattern refers to the alert indicating that the coffee maker has no water, then the system asks the user refilling the water tank.

```
<category>
<pattern>ACTION==DrinkSelection && STATE==Pause</pattern>
<action>MAKE_COFFEE</action>
<template>
<success>Making coffee</success>
<fail>Sorry, there was an error in the selection of your coffee</fail>
<template>
</category>
<category>
<pattern>QUESTION_TYPE==What && STATE</pattern>
<template>The coffee maker status is <bot state/><template>
</category>
<category>
<pattern>ALERT==NoWater</pattern>
<template>There is no coffee. Please refill the water tank. <template>
</category>
```

Fig. 1.4 An excerpt of the coffee maker's AIML-base patterns

1.4   Evaluation and Results

Despite the fact that the im4Things system is composed by several modules and services, the performed experiments were focused on evaluating the effectiveness of the present approach to control home appliances by means of written natural language instructions. In this sense, we performed a set of experiments with a first functional prototype and a set of real-world end users. This prototype has been configured for the control of five different home appliances: a coffee maker, an irrigation system, an internal lighting control, blinds, and an air heater. The overall evaluation process is described below.

1.4.1   Subjects

In order to evaluate the im4Things prototype with real-world users, the experiments were performed at the premises of the CEEIM (European Centre of Enterprises and Innovation of Murcia) located in the University of Murcia, Spain. The set of participants was composed of 5 people unrelated to the research group of which the authors are members. The age of the participants ranges from 20 to 43 years old. The participants must have two important characteristics: On the one hand, the participants must have knowledge about the use of mobile applications, particularly the use of instant messaging applications. This fact did not constitute a problem due to all participants have experience using well-known instant messaging applications such as WhatsApp® and WeChat®. On the other hand, the users must have knowledge about the services provided by the home appliances involved in the im4Things prototype. In this sense, the participants were informed of the general purpose of the study and they received an initial description of all services provided by each home appliance as well as of the im4Things mobile application.

1.4.2   Procedure

The first step of the evaluation process consisted of providing the participants a general description of the im4Things mobile application as well as of all features provided by each home appliance configured in the im4Things prototype, i.e., the coffee maker, the irrigation system, the internal lighting control, the blinds, and the air heater. After that, the im4Things mobile application was installed in the smartphones of the participants. Then, one participant was asked to add and configure all above-mentioned home appliances to the im4Things system. It is worth remembering that the im4Things mobile application allows users to interact with only one device through an individual chat window, or with more than one device through the same chat window thanks to the chatting group feature. On the basis of this understanding, the user-bot interaction performed through a chatting group with more than one im4Things bot, is the most difficult scenario from a system point of view due to the fact that, aside from the need of obtaining the command that the home appliance must execute, the system needs to determine the target home appliance. Therefore, the users were asked to create a chatting group in such a way as to enable them to control all home appliances through the same chat window, i.e., there was a total of five chatting groups each of which was integrated by one user and the five home appliances. Once all chatting groups were created, the users were asked to control the home appliances through natural language instructions. It is worth mentioning that the users used the mobile application at the premises of the CEEIM as well as from remote locations. On the other hand, due to the fact that this evaluation involves human participation, we put special attention on checking that the users were aware that these experiments were focused on evaluating the im4Things system and not the users since this fact can influence the test results [29].

Finally, once the im4Things mobile application was used by the participants along fifteen days, we analyzed all messages provided by them and used the precision, recall and F-measure metrics to obtain statistic values that allow us to determine the effectiveness of the present approach to control home appliances by means of written natural language instructions. The evaluation results are shown in the following section.

1.4.3   Results

As mentioned earlier, the present evaluation used the primary metrics precision and recall, and their harmonic mean, known as the F-measure [30]. These metrics have

Precision = True Positives / (True Positives + False Positives)
Recall =True Positives / (True Positives + False Negatives)
F-Measure = (2 * Precision * Recall) / (Precision + Recall)

Fig. 1.5 Precision, recall and F-measure formulas

Table 1.1 Evaluation results

Home appliance	Total	TP	IE	NE	Precision	Recall	F-measure
Coffee maker	200	185	10	5	0.9487	0.925	0.9367
Irrigation system	160	148	8	4	0.9487	0.925	0.9367
Internal lighting control	185	165	15	5	0.9166	0.8918	0.9041
Blinds	80	68	6	6	0.9189	0.85	0.8831
Air heater	120	106	12	2	0.8983	0.8833	0.8907
	745	672	51	22			
Average					0.9262	0.8950	0.9102

been commonly applied in information retrieval experiments [31] and natural language interfaces development [8, 26]. These metrics can be specified by using the terminology of false positives and false negatives [32]. In this context, true positives refer to the user's instructions that were successfully executed by the home appliance. The false positives refer to the user's instructions that were executed by the home appliances but they were not what user expected. Furthermore, the false negatives refer to the user's instructions that were not executed by the home appliance, but they should have been executed and the user's instructions that were executed by the home appliances but they were not what user expected. Based on the aforementioned, the precision, recall and F-measure formulas are shown in Fig. 1.5.

Table 1.1 shows the results obtained by each home appliance as well as by the whole system. From this table we can appreciate that a total of 745 instructions were provided by the group of participants, being the coffee maker, the home appliance that received the highest number of instructions. Conversely, the blinds obtained the lowest number of instructions. Also, we can see the total number of instructions that were correctly executed by the home appliance (TP—True Positives), the number of instructions that were executed by the home appliances but they were not what user expected (IE—Incorrectly executed), and the number of instructions that were not executed by the home appliance (NE—No executed). These two latter groups make up the False Negatives group.

With regard to the precision, recall and F-measure scores, the coffee maker and the irrigation system obtained the best results. Meanwhile, the blinds obtained the lowest recall score. On the other hand, the air heater obtained the lowest precision score. In general terms, the im4Things system achieved promising results with an average precision of 0.9262, a recall of 0.895, and an F-measure of 0.9102.

1.4.4   Discussion

The results obtained by the evaluation above prove that without the need for technicalities, the user was able to control the home appliances in an efficient way through the im4Things system. However, despite the results obtained seem encouraging, the im4Things system presented specific issues which are described next.

Based on a detailed analysis of the users' instructions, we ascribe the False-negatives (incorrectly executed and non-executed instructions) instructions to next reasons: (a) non-identification of keywords that help to determine the home appliance and the action to be executed, (b) user location within the home, and (c) complexity of the users' instructions. With respect to the first reason given above, some users' messages did not contain words describing the home appliance or the action to be executed, but rather they implied it. For instance, the user's message "It's too hot" does not describe the target home appliance or an action to be executed. Instead, it describes the current state of the room where the user is located. An interesting question is what else can be the system from such messages. In this sense, we are convinced that our approach can be extended to provide a more robust behavior with respect to this kind of messages. Also, we believe this improvement might be achieved through the implementation of new methods that allow us to consider a wider semantic context of the users' messages as well as the user location. Concerning user location within the home, we perceived that the main reason why the Internal lighting control got the worst results is because the users' messages were referring to a determined luminaire within the room where the user was located, e.g., "Turn off the left lamp". Based on this understanding, our approach needs to be improved through the integration of more sensors that allow the system to obtain information of the home environment and the user location within the home, thus enabling the system to take decisions based on the data obtained by the sensors. Finally, concerning the complexity of the instructions, we detected that some users' instructions required the coordination and participation of more than one home appliance, as well as of information about the current state of the room where the home appliances are located, e.g., if there are people in the room. An example of this kind of messages is "If there are not people in the main room, turn off the lights and lowers the blinds". As we can see the instruction above demands the coordination of two home appliances as well as of information about the main room. Once again, it is present the need for sensors that provide information about the room status as well as mechanisms that allow im4Things bots to cooperate among them in order to perform this kind of complex activities. From this perspective, we conclude that the im4Things system was slightly limited in dealing with complex instructions. However, we are convinced that our approach can be improved to address all above- mentioned issues without sacrificing or weakening the user expressiveness. All that previously mentioned can be achieved through the exploitation of the protocols and technologies involved in the present work. In this sense, the next section presents a detailed description of the future research.

1.5 Conclusions and Future Research

In this work, we presented a Cloud-based system to monitoring and controlling the home environment from local and remote locations. This system allows users to perform the aforementioned activities through natural language instructions by using an instant-messaging application. The natural language paradigm adopted is characterized by a flexible and intuitive way of interaction from a usage point of view. Furthermore, the user interaction paradigm adopted in this work was enhanced through the implementation of a conversational agent aiming to interact with the user through natural language, thus enhancing the user experience.

The im4Things system took advantage of technologies such as Semantic Web and the Raspberry Pi. On the one hand, the Semantic Web enabled the definition of a home appliances ontology upon which the system was able to infer the device and instruction to be executed from the natural language messages. This ontology-based approach proved to be an efficient way to process the natural language users' messages. However, the ontology needs to be extended aiming to take into account a wider set of devices available on the home environment as well as the user location. On the other hand, the Raspberry Pi technology has proved to be a smart, economic an efficient technology for implementing home appliances control through the Internet. However, it is still having small storage space and low computational efficiency. In this sense, we are convinced that this technology will be enhanced in such a way as the im4Things bot here proposed will be able to perform more demanding processes thus allowing it to become a smarter device able to cooperate with other devices to perform complex activities.

Furthermore, in this paper we presented a set of experiments performed in order to evaluate the effectiveness of the im4Things system to control home appliances by means of natural language instructions. The results obtained seem encouraging and make us believe that the system represents a solid base to face the IoT application in the home environment. However, we are aware that the system here proposed may be further expanded and improved with capabilities such as the integration of more sensors and actuators that allow to obtain information of the home environment aiming to provide suggestions to the user about energy efficiency and even allow the system itself takes certain decisions [33]. This improvement would require the extension of the home appliances ontology proposed in this work as well as of the functionalities provided by the im4Things bot. Regarding this latter issue, another potential future development refers to the communication between the im4Things bots. This feature will enable a im4Things bot to be informed about the status and configuration of other devices, thus allowing it to perform activities that involve the participation of more than one device, such as "turn off all lights that are in the same room with blinds up". It is worth noting that the correct execution of this request will provide users a complete impression of a smart home.

Finally, concerning natural language interaction, we are considering the use of emoticons in the instant messaging application. This feature will enable users to express their needs through a set of symbols in an easier and more simple way.

Despite the fact that this feature might represent an outstanding advantage, its correct implementation in the present context will demand the development of better natural language processing mechanisms.

Acknowledgements María Pilar Salas-Zárate and Mario Andrés Paredes-Valverde are supported by the National Council of Science and Technology (CONACyT) and the Mexican government.

References

1. Giusto, D., Iera, A., Morabito, G., Atzori, L. (eds.): The Internet of Things. Springer, New York (2010)
2. Dow, C.-R., Nguyen, D.-B., Wang, S.-C., Hwang, S.-F., Tsai, M.F., Dow, C.-R., Nguyen, D.-B., Wang, S.-C., Hwang, S.-F., Tsai, M.F.: A geo-aware taxi carrying management system by using location based services and zone queuing techniques on internet of things, a geo-aware taxi carrying management system by using location based services and zone queuing techniques on internet of things. Mob. Inf. Syst. Mob. Inf. Syst. **2016**, e9817374 (2016)
3. Atzori, L., Iera, A., Morabito, G.: The internet of things: a survey. Comput. Netw. **54**(15), 2787–2805 (2010)
4. Jain, S., Vaibhav, A., Goyal, L.: Raspberry Pi based interactive home automation system through e-mail. In: 2014 International Conference on Optimization, Reliabilty, and Information Technology (ICROIT), pp. 277–280 (2014)
5. Chatzigiannakis, I., Drude, J.P., Hasemann, H., Kröller, A.: Developing smart homes using the internet of things: how to demonstrate your system. In: Streitz, N., Markopoulos, P. (eds.) Distributed, Ambient, and Pervasive Interactions. Springer, pp. 415–426 (2014)
6. Zhu, J., Jia, X., Mei, X.Q.: Smart home control system based on internet of things. Appl. Mech. Mater. **738–739**, 233–237 (2015)
7. Piyare, R., Tazil, M.: Bluetooth based home automation system using cell phone. In: 2011 IEEE 15th International Symposium on Consumer Electronics (ISCE), pp. 192–195 (2011)
8. Cimiano, P., Haase, P., Heizmann, J., Mantel, M., Studer, R.: Towards portable natural language interfaces to knowledge bases—the case of the ORAKEL system. Data Knowl. Eng. **65**(2), 325–354 (2008)
9. Studer, R., Benjamins, V.R., Fensel, D.: Knowledge engineering: principles and methods. Data Knowl. Eng. **25**(1–2), 161–197 (1998)
10. Rodríguez-García, M.Á., Valencia-García, R., García-Sánchez, F., Samper-Zapater, J.J.: Ontology-based annotation and retrieval of services in the cloud. Knowl. -Based Syst. **56**, 15–25 (2014)
11. Colombo-Mendoza, L.O., Valencia-García, R., Rodríguez-González, A., Alor-Hernández, G., Samper-Zapater, J.J.: RecomMetz: a context-aware knowledge-based mobile recommender system for movie showtimes. Expert Syst. Appl. **42**(3), 1202–1222 (2015)
12. García, R., Ángel, M., Valencia García, R., Alcaraz Mármol, G., Carralero, C.: Open Idea: plataforma inteligente para gestión de ideas innovadoras. Open Idea: an intelligent platform for managing innovative ideas, September (2014)
13. Peñalver-Martinez, I., Garcia-Sanchez, F., Valencia-García, R., Rodriguez-García, M.Á., Moreno, V., Fraga, A., Sánchez-Cervantes, J.L.: Feature-based opinion mining through ontologies. Expert Syst. Appl. **41**(13), 5995–6008 (2014)
14. Katsamanis, A., Rodomagoulakis, I., Potamianos, G., Maragos, P., Tsiami, A.: Robust far-field spoken command recognition for home automation combining adaptation and multichannel processing. In: 2014 IEEE International Conference on Acoustics, Speech and Signal Processing (ICASSP), pp. 5547–5551 (2014)

15. Vanus, J., Smolon, M., Koziorek, J., Martinek, R.: Voice control of technical functions in smart home with KNX technology. In: Park, J.J.J.H., Stojmenovic, I., Jeong, H.Y., Yi, G. (eds.) Computer Science and its Applications, pp. 455–462. Springer, Berlin Heidelberg (2015)

16. Espejo Pérez, G., Ábalos Serrano, N., López-Cózar Delgado, R., Callejas Carrión, Z., Griol Barres, D.: Sistema Mayordomo: uso de un entorno de inteligencia ambiental a través de un sistema de diálogo multimodal. Mayordomo system: using an ambient intelligence environment through a multimodal dialogue system, October 2010

17. Allen, J.: Natural Language Understanding, 2nd edn. Pearson, Redwood City, Calif (1994)

18. Shelby, Z., Hartke, K., Bormann, C.: The constrained application protocol (CoAP). [Online]. Available: https://tools.ietf.org/html/rfc7252. Accessed 26 Apr 2016

19. Matsuoka, H., Wang, J., Jing, L., Zhou, Y., Wu, Y., Cheng, Z.: Development of a control system for home appliances based on BLE technique. In: 2014 IEEE International Symposium on Independent Computing (ISIC), pp. 1–5 (2014)

20. Gomez, C., Oller, J., Paradells, J.: Overview and evaluation of bluetooth low energy: an emerging low-power wireless technology. Sensors 12(9), 11734–11753 (2012)

21. Kumar, S.: Ubiquitous smart home system using android application. Int. J. Comput. Netw. Commun. 6(1), 33–43 (2014)

22. Jie, G., Bo, C., Shuai, Z., Junliang, C., Jie, G., Bo, C., Shuai, Z., Junliang, C.: Cross-platform android/iOS-based smart switch control middleware in a digital home, cross-platform android/iOS-based smart switch control middleware in a digital home. Mob. Inf. Syst. Mob. Inf. Syst. 2015(2015), e627859 (2015)

23. Smith, R.W.: Natural language interfaces. In: Encyclopedia of Language and Linguistics, 2nd edn, pp. 496–503. Elsevier, Oxford (2006)

24. Saint-Andre, P., Smith, K., Tronçon, R.: XMPP: The Definitive Guide. O'Reilly Media, Inc. (2009)

25. Alicebot and AIML Documentation (A.L.I.C.E. AI Foundation). [Online]. Available: http://www.alicebot.org/documentation/. Accessed 25 Apr 2016

26. Marston, S., Li, Z., Bandyopadhyay, S., Zhang, J., Ghalsasi, A.: Cloud computing—the business perspective. Decis. Support Syst. 51(1), 176–189 (2011)

27. Haro, L.F.D., Cordoba, R., Rojo Rivero, J.I., Diez de la Fuente, J., Avendano Peces, D., Bermudo Mera, J.M.: Low-cost speaker and language recognition systems running on a Raspberry Pi. Lat. Am. Trans. IEEE Rev. IEEE Am. Lat. 12(4), 755–763 (2014)

28. Augello, A., Pilato, G., Vassallo, G., Gaglio, S.: Chatbots as interface to ontologies. In: Gaglio, S., Re, G.L. (eds.) Advances onto the Internet of Things, pp. 285–299. Springer (2014)

29. Kaufmann, E., Bernstein, A.: Evaluating the usability of natural language query languages and interfaces to semantic web knowledge bases. Web Semant. Sci. Serv. Agents World Wide Web 8(4), 377–393 (2010)

30. Yang, Y., Liu, X.: A Re-examination of text categorization methods. In: Proceedings of the 22nd Annual International ACM SIGIR Conference on Research and Development in Information Retrieval, pp. 42–49. New York, NY, USA (1999)

31. Hripcsak, G., Rothschild, A.S.: Agreement, the F-measure, and reliability in information retrieval. J. Am. Med. Inform. Assoc. JAMIA 12(3), 296–298 (2005)

32. Advanced Data Mining Techniques. Springer, Berlin Heidelberg (2008)

33. Pilato, G., Augello, A., Gaglio, S.: Modular knowledge representation in advisor agents for situation awareness. Int. J. Semant. Comput. 5(01), 33–53 (2011)

Chapter 2
Knowledge-Based Leisure Time Recommendations in Social Networks

Dionisis Margaris, Costas Vassilakis and Panagiotis Georgiadis

Abstract We introduce a novel knowledge-based recommendation algorithm for leisure time information to be used in social networks, which enhances the state-of-the-art in this algorithm category by taking into account (a) qualitative aspects of the recommended places (restaurants, museums, tourist attractions etc.), such as price, service and atmosphere, (b) influencing factors between social network users, (c) the semantic and geographical distance between locations and (d) the semantic categorization of the places to be recommended. The combination of these features leads to more accurate and better user-targeted leisure time recommendations.

Keywords Knowledge-based recommender systems · Social networks · Collaborative filtering · Attribute constraints · Semantic information

2.1 Introduction

Knowledge-based recommender systems are a special type of recommender systems (RS) that use knowledge about users and products to pursue a knowledge-based approach to generating a recommendation, reasoning about what products meet the users' requirements [1]. Knowledge-based RS exploit semantic information to improve similarity matching between items or user profiles and items [2].

D. Margaris (✉) · P. Georgiadis
Department of Informatics and Telecommunications, University of Athens,
Athens, Greece
e-mail: margaris@di.uoa.gr

P. Georgladis
e-mail: p.georgiadis@di.uoa.gr

C. Vassilakis
Department of Informatics and Telecommunications, University of the Peloponnese,
Tripoli, Greece
e-mail: costas@uop.gr

© Springer International Publishing AG 2017   23
G. Alor-Hernández and R. Valencia-García (eds.), *Current Trends on Knowledge-Based Systems*, Intelligent Systems Reference Library 120,
DOI 10.1007/978-3-319-51905-0_2

Knowledge-based RS may employ (a) constraint-based interaction, where the user specifies constraints on the items she requests, (b) case-based interaction, where the user typically specifies specific targets and the system returns similar results and (c) may include collaborative information by identifying other users with similar profiles, and using their session information in the learning process [3].

When collaborative filtering (CF) is employed, typical RS assume that users are independent and ignore social interactions among them. Consequently, they fail to incorporate important aspects denoting interaction, such as tie strength and influence among users, which can substantially enhance recommendation quality [4, 5]. RS based on social network (SN) data tackle this issue by considering data from the user profile (e.g. location, age or gender) complemented with dynamic aspects stemming from user behavior and/or the SN state, such as user preferences, items' general acceptance and influence from social friends [4, 5]. Furthermore, tie strength between users of the SN can be exploited to further enhance the choice of recommenders, so as to consider the opinions and choices of users that have a high influence on the user for whom the recommendation is generated [6–8].

As far as SN influence is concerned, a first approach to identifying highly influential individuals within the SN is to consider those having high tie strengths, such as family members or friends with similar age; however, the influence of such individuals may be limited only to certain place categories: for instance, one may trust her friends regarding restaurants and pastry shops, but not regarding bars. Moreover, selected individuals with low tie strength, such as actors and singers, may influence a user regarding some specific categories (e.g. shops), while for some categories a user may not be influenced at all (e.g. a user may consider herself an expert in museums, hence she decides exclusively on her own, after examining the types of museums such as folk or fossil).

Recently, it has been shown that RS should consider qualitative aspects of items (typically referred to as QoS—quality of service), such as price, security, reliability, etc., the individual user behavior regarding her purchases in different item categories, and the semantic categorization of items [9, 10]. For example, if a user typically buys a glass of wine in the price range $10–$15, it would be inappropriate to recommend a restaurant charging $100 for a single glass of wine, on the grounds that some user having a high influence on the restaurant category has made a check-in in that particular restaurant. Instead, it would be more appropriate to recommend a low-cost restaurant *of the same style* with the one of the $100 glass of wine, which best fits the profile of the user to whom the recommendation is addressed.

In this chapter, we propose a novel algorithm for making accurate knowledge-based leisure time recommendations to social media users. The proposed algorithm considers qualitative attributes of the places (e.g. price, service, atmosphere), the profile and habits of the user for whom the recommendation is generated, place similarity, the physical distance of locations within which places are located, and the opinions of the user's influencers. The proposed algorithm is the first algorithm that combines the above listed features into a single and effective recommendation process, and this combination leads to increased accuracy. The proposed algorithm is evaluated both in terms of recommendation accuracy and execution performance.

In the rest of this chapter, Sect. 2.2 overviews related work, while Sect. 2.3 presents the proposed algorithm's prerequisites. Section 2.4 describes the knowledge-based recommendation algorithm for SN, while Sect. 2.5 evaluates the proposed algorithm. Finally, Sect. 2.6 concludes the chapter and outlines future work.

2.2   Related Work

Bakshy et al. [11] examine the role of SN in the RS within a field experiment that randomizes exposure to signals about friends' information and the relative role of strong and weak ties. In [12], Bakshy et al. measure social influence via social cues, demonstrate the consequences of including minimal social cues in advertising and measure the positive relationship between a consumer's response and the strength of her connection with an affiliated peer. Both these works establish that recommendation algorithms are valuable tools in SN. Oechslein et al. [7] also assert that a strong tie relationship positively influences the value of a recommendation.

In the domain of RS, numerous approaches for formulating recommendations have been proposed. Collaborative filtering (CF) formulates personalized recommendations on the basis of ratings expressed by people having similar tastes to the user for which the recommendation is generated. Taste similarity is computed by examining the resemblance of already entered ratings [13]. The CF-based recommendation approach is the most successful and widely used approach for implementing RS [14]. CF can be further distinguished in user-based and item-based approaches [15]. In user-based CF, a set of nearest neighbours of the target user is first identified, and the prediction value of items that are unknown to the target user is then computed according to this set. On the other hand, item-based CF proceeds by finding a set of similar items that are rated by different users in some similar way. Subsequently, predictions are generated for each candidate item, for example, by taking a weighted average of the active user's item ratings on these neighbour items. Item-based CF achieves prediction accuracies that are comparable to, or even better than, user-based CF algorithms [16]. To improve recommendation accuracy, knowledge-based recommender systems nowadays employ cutting-edge techniques such as data mining and segmentation [17]. Recommender systems apply to many item domains; RESYGEN [18] is a recommendation system generator that can generate multi-domain systems. For computing similarities in the recommendation process, RESYGEN provides a similarity metrics library and the RS configurator chooses the most appropriate one.

Recently, SN recommendation has received considerable research attention. Konstas et al. [19] investigate the role of SN relationships in developing a track

recommendation system using CF and taking into account both the social annotations and friendships inherent in the social graph established among users, items
and tags. Arazy et al. [6] outline a conceptual RS design within which the structure
and dynamics of a SN contribute to the dimensions of trust propagation, source's
reputation and tie strength between users, which are then taken into account to
generate recommendations. Quijano-Sanchez et al. [8] enhance a content-based RS
by considering the trust between individuals, users' interaction and aspects of each
user's personality. In [20], a matrix factorization-based approach for recommendation in SN is explored, employing a mechanism of trust propagation. He et al. [5]
analyze data from a SN and establish that friends have a tendency to select the same
items and give similar ratings; they also show that using SN data within the RS
improves prediction accuracy but also remedies the data sparsity and cold-start
issues inherent in CF.

As far as leisure time place recommendation is concerned, Zheng et al. [21]
make personalized travel recommendations from user GPS traces, by modeling
multiple users' location histories and mining the top n interesting locations and the
top m classical travel sequences in a region. Bao et al. [22] present a location-based
and preference-aware RS, which provides a user with location recommendations
around the specified geo-position based on (a) the user's personal preferences learnt
from her location history and (b) social opinions mined from the *local experts* who
could share similar interests. RecomMetz [23] is a context-aware knowledge-based
mobile RS specialized in the domain of movie showtimes based on location, time
and crowd information. RecomMetz views recommended items as composite ones,
with salient aspects being the theatre, the movie and the showtime. iTravel [24] is
an attraction recommendation, employing mobile peer-to-peer communications for
exchanging ratings via users' mobile devices, and using these ratings for recommendation formulation. Moreno et al. [25] present SigTur/E-Destination, a RS for
tourist and leisure activities situated in the region of Tarragona, Spain.
SigTur/E-Destination exploits semantic information attached to leisure activities
and opinions from similar users to generate recommendations. Ference et al. [26]
study the issues in making location recommendations for out-of-town users, by
taking into account user preference, social influence and geographical proximity
and introduce UPS-CF, a recommender engine for making location recommendation for mobile users in location-based SN such as FourSquare. Table 2.1 depicts a
feature comparison between the presented algorithm and other leisure time place
recommenders surveyed in this section.

In Table 2.1, we can see that the proposed algorithm is the only one supporting
QoS aspects, while it additionally supports all other dimensions (QoS, SN, semantic
matching and proximity). Furthermore, the proposed algorithm computes influence
among user pairs in SN per interest category (as contrasted to computations only at
user level) and uses semantic distances between (a) places and (b) the locations
within which the places are located. These features, unique to the proposed algorithm, enable the formulation of highly accurate recommendations.

Table 2.1 Leisure time place recommenders feature comparison

Algorithm-reference	Recommendation domain	QoS aspects?	SN-aware?	Semantics?	Proximity?
Zheng et al. [21]	Locations, travel sequences	No	No	No	Yes
Bao et al. [22]	Nearby locations	No	Yes	No	Yes
RecomMetz [23]	Movie showtimes	No	No	Yes	Yes
iTravel [24]	Attractions	No	Yes	No	Yes
SigTur/E-Destination [25]	Leisure activities in Taragona region	No	No	Activities	Yes
UPS-CF [26]	Locations	No	Yes	No	Yes
Proposed algorithm	Places	Yes	Yes	Yes	Yes

2.3   Social Networking, Semantics and QoS Foundations

In this section, we summarize the concepts and foundations from the areas of SN, semantic data management and quality of service (QoS), which are used in this work. QoS, in particular, relates to important attributes on which constraints are imposed within the knowledge-based recommendation process, hence it plays a central role in the presented knowledge-based recommendation algorithm.

2.3.1   Influence in Social Networks

Within a SN, "social friends" greatly vary regarding the nature of the relationship holding among them: they may be friends or strangers, with little or nothing in between [27]. Users have friends they consider very close, and know each other in real life and acquaintances they barely know, such as singers, actors and athletes [28]. Bakshy et al. [12] suggest that a SN user responds significantly better to recommendations (e.g. advertisements) that originate from friends of the SN to which the user has a high *tie strength*. In their work, the strength of the directed tie between users i and j is linked to the amount of communication that has taken place between the users in the recent past and is computed as:

$$W_{i,j} = \frac{C_{i,j}}{C_i} \tag{2.1}$$

where C_i is the total number of communications posted by user i in a certain time period (a period of 90 days is considered for computing the tie strength) in the SN, whereas $C_{i,j}$ is the total number of communications posted on the SN by user i during the same period and are directed towards user j or on posts by user j.

Although the tie strength metric can be used to locate the influencers of a user, it does not consider user interests, which are important in RS. In our work, we consider a more elaborate influence metric, which computes the tie strength between users i and j for each distinct interest. In more detail, the influence level $IL_{i,j}(C)$, where C is an interest category is defined as follows:

$$IL_{i,j}(C) = \begin{cases} W_{i,j}, & \text{if } C \in interests(i) \wedge C \in interests(j) \\ 0, & \text{otherwise} \end{cases} \qquad (2.2)$$

Effectively, this formula assigns a zero influence level value for interests that are not shared among the considered users, whereas for common interests, the value of the tie strength is used. For the population of each user's interest set, we use the user interest lists collected by the SN [29]. Since this list is built automatically when the user interacts with the SN, it will be comprehensive and will include all categories that the user is interested in.

2.3.2 Leisure Time Places Semantic Information and Similarity

To generate successful recommendations, the algorithm must be able to find which leisure places are similar. This is achieved through recording semantic information related to the places and using this information to compute semantic similarity among them. Semantic information is stored in ontologies and Fig 2.1. illustrates an ontology concerning places: the "Building" entity is the root of the *is-a* hierarchy, and it is subclassed to generate more specific place categories such as attraction, accommodation, leisure. Each class/subclass is described through a set of properties which apply to all instances of the particular class as well as to instances of its subclasses (e.g. an attraction may have a property expressing age suitability, which is inherited to all its subclasses, namely religious monuments and museums).

In this work, we adopt a modified version of the similarity measure proposed by [30]: the semantic similarity (*SemSim*) between two places p_i and p_j is based on the ratio of the shared Resource Description Frameworks (RDF) descriptions between p_i and p_j (*count_common_desc*(p_j, p_j)) to their total descriptions (*count_total_desc* (p_i, p_j)), i.e.:

$$SemSim(p_i, p_j) = \frac{count\_common\_desc(p_i, p_j)}{count\_total\_desc(p_i, p_j)} \qquad (2.3)$$

However, the RDF descriptions of two places may not be identical, yet be semantically close: e.g., a restaurant with Lebanese cuisine can be considered of high similarity to a restaurant with Tunisian cuisine, but of low similarity with a Japanese restaurant. To address this aspect, we modify the similarity metric formula to

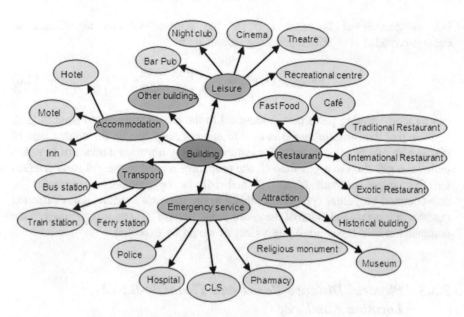

Fig. 2.1 Example of leisure time ontology tree

$$SemSim(p_i, p_j) = \frac{\sum_{p \in p_i \land p \in p_j} sim_p(V_p(p_i), V_p(p_j))}{count\_total\_desc(p_i, p_j)} \qquad (2.4)$$

where p is a property, $V_p(p_i)$ and $V_p(p_j)$ are the values of property p for items p_i and p_j, respectively, and sim_p is a function computing the similarity between values of property p. In the example given above, we may consider that $sim_{cuisine}(Lebanese, Tunisian) = 0.9$ (i.e. a high value) and $sim_{cuisine}(Lebanese, Japanese) = 0.1$ (i.e. a low value). For many properties with a numeric (integer or real) domain (e.g. prices, distances, hotel star ratings etc.) the sim_p function may be defined as

$$sim_{num\_prop}(v1, v2) = 1 - \frac{|v1 - v2|}{\max(num\_prop) - \min(num\_prop)} \qquad (2.5)$$

where $max(num\_prop)$ and $min(num\_prop)$ are the maximum and minimum values respectively of $numeric\_prop$ in the ontology extension; this is a typical value normalization formula [31]. Undoubtedly, defining a similarity function for each property within ontology is a laborious task. To mitigate this issue, automated similarity computation methods for specific domains can be used. For instance, metrics sim_g and sim_d [32] can be used for movie genres and movie directors; the metrics in [33] can be used for colors; and so forth. To further decrease the amount of similarity functions that need to be defined, place similarity computation may consider only the places' salient features (e.g. for a museum, the ticket cost and the type of its collections (such as gallery, antiquities, modern art) are salient features, but its number of floors is not). In the absence of any algorithmic or custom-provided similarity metric, the default metric

$$sim_{default}(v1, v2) = \begin{cases} 1, & \text{if } v1 = v2 \\ 0, & \text{otherwise} \end{cases} \qquad (2.6)$$

can be used, offering performance identical to the one of the method used in [30].

Note that the *SemSim* metric is able to appropriately handle the comparison of places with overlapping features: for instance, when comparing a museum M which includes a gift shop to a gift shop G, the properties related to the gift shop features will be present in both M and G and will be appropriately compared. The non-common properties (properties related to the museum's nature, e.g. *Collection types*) will increase the value of the *count_total_desc*(M, G) quantity, leading to the computation of a lower similarity value, as would be expected.

2.3.3 Physical Distance-Based and Thematic-Based Location Similarity

Besides the semantic information considering places, the physical distance and the thematic similarity of the locations within which the places are located have been proved to play an important role when humans search for places that include a geographical dimension [34]. According to Jones et al. [34], the following criteria can be used to assess the physical distance-based similarity: (a) distance in map or geographical coordinate space between locations, (b) travel time between locations, (c) number of intervening places, (d) spatial inclusion of one location in the other, (e) overlapping between locations, and (f) boundary connectivity between locations.

For physical distance-based similarity, Jones et al. [34] compute a combined spatial closeness measure, called *Total Spatial Distance* (*TSD*) using the formula:

$$TSD(loc_1, loc_2) = w_e * ED(loc_1, loc_2) + w_h * HD(loc_1, loc_2) \qquad (2.7)$$

where *ED* is the (normalized) Euclidian distance between the two locations and *HD* is the respective hierarchical distance (again, normalized), which is computed using an hierarchy of part-of relations (e.g. Paris is *part-of* Île-de-France, which is *part-of* France etc.). w_e and w_h are weights assigned to *ED* and *HD* respectively; in Jones et al. [34], w_e is set to 0.6 and w_h to 0.4. Metric value normalization is performed using a formula analogous to the sim_{num_prop} formula (c.f. Sect. 2.3.2).

Regarding the thematic-based location similarity, Jones et al. [34] introduce a *Thematic Distance* (*TD*) metric, which takes into account the semantic similarity of classification terms attached to each location; these terms are drawn from the Art and Architecture Thesaurus taxonomy. Locations having attached classification terms that are semantically close have small thematic distance, while locations having attached semantically dissimilar classification terms have a high thematic distance.

Finally, in order to combine *TSD* and *TD* into a single score, Jones et al. [34] employ the weighted average formula shown in Eq. (2.8), setting the weight w_t of *TD* to 0.4 and the weight w_s of *TSD* to 0.6. In our work, we have adopted the approach of [34], modified to allow the use of arbitrary classification terms instead of terms drawn from a specific taxonomy; to compute the semantic distance between classification terms attached to places, we used the *word2vec* library [35].

$$LocationSim(loc_1, loc_2) = 1 - (w_t * TD(loc_1, loc_2) + w_s * TSD(loc_1, loc_2)) \quad (2.8)$$

2.3.4 Leisure Time Places QoS Information

QoS may be defined in terms of attributes [36]. Attributes typically considered in the context of QoS are cost, timeliness, reliability, courtesy, etc. [37]. In this chapter we will consider only the attributes cost (c), service (s) and atmosphere (a), as used in many of the travel websites, such as Tripadvisor or Opentable. When choosing a place to visit, users typically try to minimize cost and maximize service and atmosphere. It is straightforward to extend the algorithm presented below for handling QoS information to include more attributes, hence the consideration of only three attributes does not lead to loss of generality. Regarding the cost, actual prices are used, while values for the service and atmosphere scores are taken from sites such as TripAdvisor (e.g. http://www.tripadvisor.com/Restaurants-g186338-London_England.html refers to London restaurants). These values are normalized in the scale of 1–10, with larger values denoting higher quality. An example of the London's restaurants qualitative characteristics values are shown in Table 2.2.

2.3.5 User's Profile for Enabling Recommendations

As discussed in the introduction, users are influenced regarding leisure time places they visit by other users; the set of influencers may vary between place categories, e.g. user u may trust her friends f_1 and f_2 regarding restaurants, but regarding bars

Table 2.2 Sample QoS values within the repository

Place	Cost $	Service	Atmosphere
Restaurant Gordon Ramsay	140	10	9
Italian Pizza Connection	350	9	8
London Fish & Chips	8	8	7
...			

she may trust her friend f_3 and be influenced by the choices made by actor A. Moreover, a user u may make decisions on her own for a particular place category, by personally locating candidate places and examining their characteristics. In order to accommodate these aspects in the RS, we follow the approach presented in [38], adapting it appropriately. According to this approach, in order to formulate a leisure time place recommendation, two subtasks are executed in parallel: the first task computes a QoS-based recommendation, while the second task computes a CF-based recommendation. Then, the two recommendations are combined to formulate the final recommendation, employing a metasearch algorithm [31].

In our case, the CF-based algorithm considers the opinions of the user's influencers for the particular leisure time place category. A distinct set of influencers is maintained for each place category, to increase the accuracy of the recommendations, and the sets of influencers per place category are maintained in the user profile. The QoS-based algorithm considers only the qualitative characteristics of each place. Additionally, for each user we store in her profile the average values of the QoS attributes (cost, service and atmosphere) of the places she visits for different places categories (museums, bars, etc.). This enables us to determine how close each place is to the visiting habits of the particular user.

In order to combine the QoS-based recommendation and the CF-based recommendation into a single recommendation for the user, we use the $WCombSUM_i$ formula [39]. According to this formula, the overall score for an item i within the final recommendation for user u is

$$WCombSUM_{i,u} = w_{CF,C(i),u} * score_{CF,i,u} + w_{QoS,C(i),u} * score_{QoS,i,u} \qquad (2.9)$$

where $score_{CF,i,u}$ and $score_{QoS,i,u}$ are the recommendation scores for item i produced for u by the CF-based and the QoS-based algorithm respectively and $C(i)$ denotes the category of item i. $w_{CF,C(i),u}$ and $w_{QoS,C(i),u}$ are weights assigned to the CF-based and the QoS-based algorithm respectively. In order to provide highly personalized recommendations, algorithm weights are computed individually for each user and category, i.e. two distinct users may have different weights for the same category, while different weights may apply for a particular user u when considering distinct categories c_1 and c_2, (e.g. "museums" and "restaurants"). Weight values are computed using the following formulas:

$$w_{CF,C(i),u} = \frac{|PlacesVisited_{C(i),u} \cap PlacesVisitedByInfluencers_{C(i),u}|}{|PlacesVisitedByInfluencers_{C(i),u}|} \qquad (2.10)$$

$$w_{QoS,C(i),u} = 1 - w_{CF,C(i),u}$$

Effectively, $w_{CF,C(i),u}$ is the ratio of the places visited by u within C_i and have been recommended by influencers to the overall number of places visited by u within C_i. Clearly, the higher this ratio, the more receptive u is to suggestions made by influencers, hence the weight assigned to the CF-based algorithm increases.

Regarding the calculation of the set $PlacesVisitedByInfluencers_{C(i),u}$, a visit is considered to have been suggested by an influencer if (a) an influencer has visited the same place two years or less prior to the user's visit, (b) the user had not visited the place before the considered influencer and (c) a recommendation had been made to the user, triggered by the influencer's visit. The time frame of two years has been chosen so as to (i) include in the "influence window" two consecutive similar tourist seasons (e.g. a visit made by a user to a summer resort in August 2015 is considered to have been influenced by an influencer's visit to the same resort made in June 2014) and (ii) allow for the decaying of information that was collected long ago.

The formula computing the CF algorithm weight suffers from the cold start problem, i.e. the case that no (or very few) data are present in the system for a specific category. In more detail, if no places of a specific category have been visited by influencers, the formula is not computable; additionally, if the number of places that have been visited by influencers, within a specific category, is small, the result computed by the formula will not be indicative of how receptive a user is to her influencers' suggestions (due to lack of statistical significance). Therefore, when the cardinality of the $PlacesVisitedByInfluencers_{C(i),u}$ set is below a threshold th, we set $w_{CF,C(i),u}$ to a default value of 0.4. The value of 0.4 has been chosen based on the work presented in [38], which asserts that a value equal to 0.4 ensures that the recommendations adhere to the QoS levels desired by the user, while at the same time the opinions of the influencers have an adequately strong effect in the formulation of the final recommendation. In our work, we have used a value of th equal to 10, since this has been experimentally proven to be a sufficient number of elements to generate an acceptably accurate value for $w_{CF,C(i),u}$.

2.4   The Leisure Time Recommendation Algorithm

Having available the information listed in Sect. 2.3, the algorithm performs the three steps listed below.

Step 1—Offline Initialization. The algorithm is initially bootstrapped by executing the following actions:

- for each place category C, the minimum and maximum place cost in the category are identified, using the formulas shown in Eq. (2.12). Similarly, the $minSer(C)$ and $maxSer(C)$ and $minAtm(C)$ and $maxAtm(C)$ quantities are computed, corresponding to the minimum and maximum service and atmosphere of places in category C, respectively.

$$minCost(C) = \min_{place_i \in C} (cost(place_i))$$
$$maxCost(C) = \max_{place_i \in C} (cost(place_i))$$

(2.11)

- for each user u and place category C, the algorithm computes the values of $w_{CF,C(i),u}$ and $w_{QoS,C(i),u}$ according to the formulas presented in Sect. 2.3.5, and the mean cost, service and atmosphere of the places within category C that user u has visited in the past.
- for each user u and place category C, the algorithm computes the influence level of her social friends (c.f. Sect. 2.3.1), and then retains the *top* N social friends with the highest influence level in C. In this work, we set $N = 6$, since we have experimentally determined that this value is adequate for producing accurate recommendations; this experiment is described in Sect. 2.5.1.
- for each pair of places (p_1, p_2), the recommendation algorithm computes the place similarity between p_1 and p_2. The place similarity metric takes into account both the semantic similarity between the places and the similarity between the locations within which the places are located and is computed as

$$PlaceSim(p_1, p_2) = SemSim(p_1, p_2) * LocSim(loc(p_1), loc(p_2)) \qquad (2.12)$$

where *SemSim* (p_i, p_j) is the semantic similarity between places p_i and p_j (c.f. Sect. 2.3.2, $loc(p)$ denotes the location in which p is located, and *LocSim* (loc_i, loc_j) denotes the similarity of locations loc_i and loc_j. Recall than the *LocSim* metric encompasses both the physical and the thematic distance between the locations (c.f. Sect. 2.3.3).

Step 2—Online operation: After the initialization phase, the algorithm can be executed in an online fashion to produce recommendations. The algorithm's execution is triggered by events generated by users: in particular, the algorithm considers the generation of recommendations each time a user checks-in a leisure time place or is tagged to be in some leisure time place.

When a user *infl* checks-in or is tagged to be in a leisure time place p that belongs in category C, the algorithm considers to make a recommendation to those users that are influenced by *infl* on C. To this end, the algorithm computes the set *PRR(C, infl)* of Potential Recommendation Recipients as follows

$$PRR(C, infl) = \{u | infl \in influencers(u, C)\} \qquad (2.13)$$

Subsequently, for each user u in *PRR* $(C, infl)$ the algorithm computes which recommendation should be sent to the particular user. The rationale to formulate the recommendation to be sent is as follows:

- If the QoS parameters of p are "close" to the QoS attributes of places that user u typically visits within C, then the algorithm checks if p might be of interest to the user, considering the opinion of u's influencers in C and the QoS parameters of p. If the algorithm determines that p might be of interest to user u, then p is recommended to u.

- If the QoS parameters of p are too distant from the QoS attributes of places that user u typically visits within C, then the algorithm searches for a place p' in C that (a) is highly similar to p and (b) its QoS attributes are close to the QoS attributes of places that u typically visits within C. Then, the algorithm checks if p' could be of interest to u, considering the opinion of u's influencers within C, the similarity between p and p' and the QoS parameters of p'. If the algorithm determines that p' might be of interest to u, then p' is recommended to u.

In more detail, initially, the QoS score of place p for user u is computed as follows:

$$score_{QoS,p,u} = cost\_vicin(u,p) * ser\_vicin(u,p) * atm\_vicin(u,p) \qquad (2.14)$$

where

$$cost\_vicin(u,p) = 1 - \frac{|cost(p) - MC(u,C)|}{maxCost(C) - minCost(C)}$$

$$ser\_vicin(u,p) = \begin{cases} 1 - \frac{|ser(p) - MS(u,C)|}{maxSer(C) - minSer(C)}, & if\ ser(p) \le MS(u,C) \\ 1, & if\ ser(p) > MS(u,C) \end{cases} \qquad (2.15)$$

$$atm\_vicin(u,p) = \begin{cases} 1 - \frac{|atm(p) - MA(u,C)|}{maxAtm(C) - minAtm(C)}, & if\ atm(p) \le MA(u,C) \\ 1, & if\ atm(p) > MA(u,C) \end{cases}$$

In the equations above, $cost(p)$ is the average cost (entrance ticket or cost of items that are sold there) of p, $ser(p)$ and $atm(p)$ are the service atmosphere ratings of p (all values are copied from sites such as www.tripadvisor.com), while $MC(u, C)$, $MS(u, C)$ and $MA (u, C)$ are the mean cost, mean service and mean atmosphere respectively of places visited by u within C. Cost vicinity indicates how close the place price is to the user's price habits within the specific category.

When computing service vicinity or atmosphere vicinity, we consider a place close to the user's preferences if its service is either equal to or surpassing the mean service (respectively atmosphere) rating of the places that the user visits in this category. The rationale behind this calculation is that places having received high marks for these criteria would typically be of higher interest to the user. For places having service less than $MS(u, C)$ [respectively, atmosphere less than $MA(u, C)$], a typical normalized distance metric [31] is employed.

If $score_{QoS,i,u}$ is greater than 0.68 (a discussion on the choice of this value is given in the conclusions section), then the QoS parameters of p are considered to be close to the QoS attributes of places that u typically visits within C. In this case, the algorithm computes the CF-based score for the recommendation of p: it first extracts the N influencers regarding C from the u's profile and, for each of these influencers IN locates the place p_{IN} within C that she has visited and has the greatest place similarity with p. Then, the CF-based score $score_{CF,p,u}$ is computed as

$$score_{CF,p,u} = \frac{\sum_{IN \in influencers(u,C)} *PlaceSim(p,p_{IN})}{\sum_{IN \in influencers(u,C)} IL_{u,IN}(C)} \tag{2.16}$$

Finally, the probability that u is interested in p is computed as follows:

$$IP_{u,p} = w_{CF,C(p),u} * score_{CF,p,u} + w_{QoS,C(p),u} * score_{QoS,p,u} \tag{2.17}$$

where $w_{CF,C(p),u}$ and are $w_{QoS,C(p),u}$ are the weights assigned to the CF and QoS dimension regarding recommendations made to u for category $C(p)$ (i.e. the category of the place appearing in the event—for details on the computation of these weights, c.f. Sect. 2.3.5). If the value of $IP_{u,p}$ meets or exceeds the interest probability (IP) threshold (Sect. 2.5.3 discusses the computation of the IP threshold's value), p is recommended to user u.

If $score_{QoS,i,u}$ is less than 0.68, the QoS levels of p are considered distant from those that u typically visits in C. In this case, the algorithm searches to locate a place p' within C that is similar to p and has QoS levels close to the habits of u. This is achieved by finding the place with the maximum value

$$PlaceSim(p,p') * score_{QoS,p',u} \tag{2.18}$$

For p', the IP metric is computed as described above, and if it is higher than the IP threshold, p' is recommended to u.

It is worth noting that in the absence of a history of visits made by a user (i.e. a new user or a user that recently visited her first place in this particular category), the quantities MC (u, C), MS (u, C) and MA (u, C) used for the computation of the QoS-based score are computed taking into account the relevant quantities of the user's influencers in the particular category. More specifically MC (u, C) is computed as

$$MC(u,C) = \frac{\sum_{IN \in influencers(u,C)} IL_{u,IN}(C) * MC(IN,C)}{\sum_{IN \in influencers(u,C)} IL_{u,IN}(C)} \tag{2.19}$$

where $IL_{u,IN}(C)$ is the influence level of user IN on user u regarding category C (c.f. Sect. 2.3.1). Analogous computations are performed for MS (u, C) and MA (u, C).

Step 3—Repository update. Since both the content of the social networks and the places information are dynamic, a number of information elements of our model needs to be updated, in order to maintain consistency with the current status of the social network and leisure time place information. The updates that need to be performed are as follows:

1. Each time a new place is introduced, a check needs to be made whether the minimum and maximum values of the QoS parameters within that category need to be updated.

2. After a user checks-in a place or is tagged to be there, the profile of the user is updated regarding the mean QoS attributes (cost, service and atmosphere) of the places category that the place belongs to.
3. The weights assigned to the CF dimension ($w_{CF,C(p),u}$) and the QoS-based dimension ($w_{QoS,C(p),u}$) need to be recomputed when the underlying data (places that a user has checked-in or has been tagged to be there, and visits to places suggested by influencers) change.
4. Each time a new place is introduced, place similarities between the new place and all places in the database are also computed.
5. Finally, the top influencers of each user u within each category of interest C need to be recomputed when the underlying data (categories of interest and/or number of communications) change.

Updates (1) and (2) include only basic computations hence they can be performed synchronously with the triggering event. Updates (3)–(5) are more computationally expensive and can be performed in an offline fashion, e.g. be executed periodically.

2.5 Experimental Evaluation

In this section, we report on our experiments through which:

1. We determined the values of parameters that are used in the algorithm. More specifically, an initial experiment was conducted to evaluate the optimal value for parameter N, expressing the number of influencers that must be maintained per place category so as to provide accurate recommendations. A second experiment was targeted to estimate the taxonomy level of categories that should be retained within the profile of each user (i.e. the taxonomy level of place categories for which we maintain the weights Cf_{weight} and QoS_{weight}, as well as the top influencers for each user), in order to assess the storage requirements and the scalability of the approach.
2. We evaluated the performance of the proposed approach, both in terms of execution time (the time needed to make the recommendations) and users' satisfaction regarding the offered recommendations.

For our experiments we used two machines. The first was equipped with one 6-core Intel Xeon E5-2620@2.0GHz CPU and 16 GB of RAM, which hosted the processes corresponding to the active users (browser emulators), i.e. the users who generated the triggering events. The second machine's configuration was identical to the first, except for the memory which was 64 GB; this machine hosted (i) the algorithm's executable, (ii) a database containing the users' profiles including the influence metrics per category, the lists of top N influencers per category and the data regarding the tags and check-ins made by each user and (iii) the places database, which includes their semantic information and QoS data (cost, service and atmosphere). The machines were connected through a 1 Gbps local area network.

To assess recommendation quality, we conducted a user survey in which 60 people participated. The participants were students and staff from the University of Athens community, coming from 4 different academic departments (computer science, medicine, physics and theatre studies). 29 of the participants were women and 31 were men, and their ages range between 18 and 48 years old, with a mean of 29. All of the participants were Facebook users, and we extracted the profile data needed for the algorithm operation using the Facebook Graph API (https://developers.facebook.com/docs/graph-api). Regarding the participants' profile and behavior within Facebook, the minimum number of Facebook friends among the participants was 148 and the maximum was 629 (with a mean of 264). All participants used Facebook at least 4 days a week and one hour per day of use, and had been members for at least two years. For each person, we computed the relevant tie strengths with all of her Facebook friends in an offline fashion.

The places data used in the experiment were extracted from Tripadvisor. The data set consisted of 5000 places in 20 cities (including New York, Los Angeles, London, Rome, Paris, Dubai, Athens and Beijing) and falling in 10 places categories (museums, religious/historical monuments, bars, nightclubs, cinemas, theatres, fast food restaurants, cafés and restaurants). The cost attribute values in this repository were set according to the places' current prices, while the service and atmosphere attribute values were set according to the users' rating summary from Tripadvisor.

2.5.1   Determining the Number of Influencers

The first experiment aimed to determine the number of influencers N that must be maintained by the system per user and per category, in order to produce accurate recommendations; it is desirable that this number is kept to a minimum, to save space and limit the amount of data to be processed, thus reducing recommendation time. Recall that for each place category, the algorithm considers the opinions of the strongest influencers within the specific category when recommendations are generated. To find the minimum number of influencers that can be kept without limiting the quality of recommendations, we gradually increased the number of strongest influencers maintained, seeking the point at which considering more influencers does not modify the generated recommendations. The generated recommendations will converge when the number of considered influencers increases, because stronger influencers are added first to the set of considered influencers, hence increments beyond some point will only lead to incorporation of weaker influencers, and this incorporation will not result to modification of the generated recommendation.

To identify the value of N after which recommendations remain stable, we generated 1000 synthetic (*user, category*) pairs to formulate recommendations for.

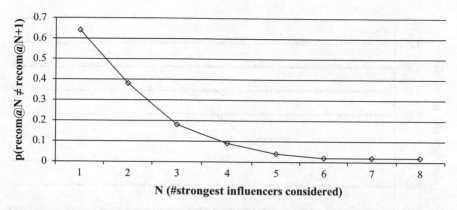

Fig. 2.2 Different recommendations made, due to the fact of considering 1 more recommender

Subsequently, we iterated on the generation of the recommendations, for different values of N varying from 1 to 10. Finally, we calculated the probability that a recommendation generated for a user considering her N strongest influencers (*recom@N*) is different than the corresponding recommendation generated considering her $N+1$ strongest influencers (*recom@N+1*); this probability is denoted as $p(recom@N{\neq}recom@N+1)$. The results, depicted in Fig. 2.2, show that any further increments beyond the number of 6 influencers result only in marginal modifications to the recommendations (98% of the recommendations remain unmodified). Therefore, in the following experiments we fix the number of N to 6.

2.5.2   Estimating the Taxonomy Level of Places Categories of Interest per User

To estimate the taxonomy level of places categories of interest per user, we conducted an experiment with our 60 participants. In this experiment, we varied the taxonomy level maintained in the profiles using the following values: level 0 (average of all the places, in general), level 1 (taxonomy level examples: leisure, restaurant, attraction, etc.), level 2 (examples: nightclubs, cinemas, fast food restaurants, cafés, museums, etc.) and level 3 (examples: Asian restaurants, operas, internet cafés, folk museums, etc.). Then, for each setting, 10 recommendations were generated for each user, and the user was asked to assess the probability that she would visit the recommended place.

The results, depicted in Fig. 2.3, show that when taxonomy is considered at level 0, visiting probability is very low (3%). Considering the taxonomy at level 1 raises visiting probability to 14%, while considering levels 2 and 3 raises further visiting probability to 30 and 35%, respectively.

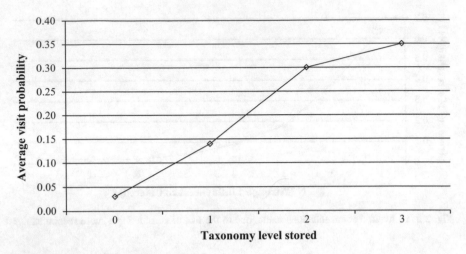

Fig. 2.3 Visit probability when information is stored at different taxonomy levels

Using the fourth level of the taxonomy (actual places) was not considered, since in this case the algorithm could produce only few useful recommendations: recall the algorithm recommends to users places in categories they are interested in, with interest category population relying on the interest lists collected by the social network. Therefore, for a place to be recommended the particular place would have to be included in such an interest list, which would typically require that the user has either visited the place's profile in the social network or has engaged in some activity or conversation about it and thus is already aware of its existence (and thus the recommendation offers little or no new information).

Regarding the data size, if we choose to maintain user preferences at taxonomy level-2, we need to store approximately 4 KB per user. However, if we choose to maintain user preferences at taxonomy level-3, storage requirements raise to approximately 100 KB per user (in the worst case scenario, where our user has an interest in all the 2000 kinds of places; the number of 2000 corresponds to the count of level-3 taxonomy branches in the SN user preference database). Hence, in the subsequent experiments we will store our data in taxonomy level-3 detail, since we can achieve better accuracy and—at the same time—the related storage requirements can be handled by the current technology.

2.5.3 Interest Probability Threshold

As discussed in Sect. 2.4, the algorithm uses an interest probability threshold to decide whether some recommendation is of interest to the user. To determine the threshold, we exploited the data gathered in the experiment described in Sect. 2.5.2. Figure 2.4 displays the relation of the interest probability metric (IP) introduced in

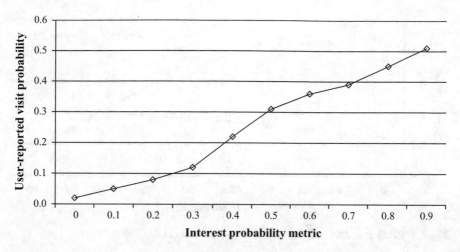

Fig. 2.4 Visit probability against interest probability metric

Sect. 2.4 to the user-reported probability for following the recommendation. From this figure, we can determine that there is positive correlation between the IP metric and the probability that the user actually follows the recommendation. Clearly, recommendations with a high IP metric are likely to be followed by users (and should thus be forwarded to them) while those with low IP metric will probably be ignored (and should thus be suppressed).

The optimal setting for the IP threshold would (a) maximize the probability that a forwarded recommendation is useful (i.e. it would be followed) and (b) maximize the probability that a recommendation that would be followed is not suppressed. In a formal notation, item (a) corresponds to maximizing the quantity

$$Prec = \frac{|\{useful\_recommendations\} \cap \{forwarded\_recommendations\}|}{|\{forwarded\_recommendations\}|} \quad (2.20)$$

while item (b) corresponds to the maximization of the quantity

$$Rec = \frac{|\{useful\_recommendations\} \cap \{forwarded\_recommendations\}|}{|\{useful\_recommendations\}|} \quad (2.21)$$

These goals are contradicting: goal (a) suggests that a high threshold is used, so that only the recommendations of high interest (and thus of a high probability to be followed) are forwarded to the user. On the contrary, goal (b) suggests that a low threshold is used, so that users are not deprived of recommendations that they would follow, even if the relevant IP metric value is small. Figure 2.5 illustrates this contradicting relation of the *Prec* and *Rec* metrics to the IP threshold.

To best serve both these contradicting goals, we seek to maximize their harmonic mean, i.e. the quantity

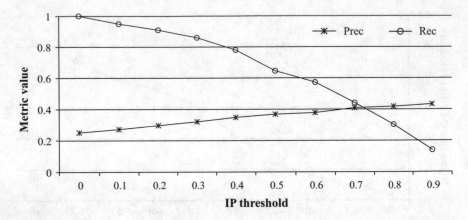

Fig. 2.5 Relation of *Prec* and *Rec* metrics to the IP threshold

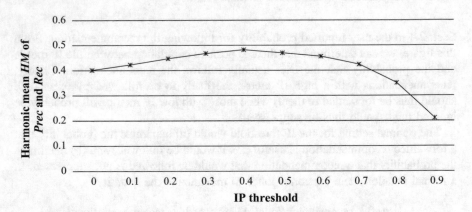

Fig. 2.6 Relation of the harmonic mean of *Prec* and *Rec* metrics to the IP threshold

$$HM = 2 * \frac{Prec * Rec}{Prec + Rec} \qquad (2.22)$$

analogously to the combination of the precision and recall metrics in information retrieval, which produces the F1-measure [40]. As shown in Fig. 2.6, the value of the IP threshold that maximizes the harmonic mean is 0.4; hence, in the following experiments we will set the IP threshold to this value.

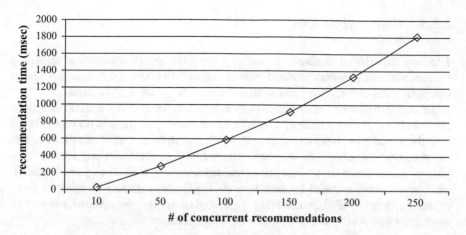

Fig. 2.7 Time needed for recommendation formulation

2.5.4   Recommendation Formulation Time

The next experiment is aimed at measuring the time needed for recommendation formulation when a triggering event occurs (users check-in or are tagged). To measure the time needed, we created a synthetic user base consisting of 50,000 users. Each user had from 100 to 1000 friends overall, with an average of 190 friends, following the mean value of friends on Facebook (https://web. facebook.com/notes/facebook-data-team/anatomy-of-facebook/101503885192438-59?_rdr). Each user was set to have a history of 0–100 visits. All repositories (the places' semantic repository, the places' qualitative repository, the users' top recommenders and each user's past visits) were implemented as in-memory hash-based structures, which proved more efficient than using a separate database, such as HSQLDB (http://hsqldb.org/) (memory-based) or MySQL (http://www.mysql.com) (disk-based).

The measurements obtained from this experiment are depicted in Fig. 2.7. We can observe that the time needed remains low even for high degrees of concurrency (approximately 0.6 s for 100 concurrent recommendations) and scales linearly with the concurrency degree. Even when 250 recommendations need to be concurrently generated (250 users have simultaneously checked-in or been tagged), the average recommendation time is less than 2 s, which is deemed satisfactory for the infrastructure capacity. Note that the corresponding Facebook infrastructure was estimated to over 60,000 servers for approximately 608 million users in June 2010 [41], which gives about 10,000 users per server; in our experiment, a mid-range workstation was set to handle 50,000 users.

44

2.5.5 User Satisfaction

The final experiment is aimed at assessing the participants' satisfaction regarding the recommendations they received from the algorithm presented in Sect. 2.4, and at comparing this satisfaction level to that obtained from other algorithms.

In this experiment, each participant was asked to rate 40 recommendations presented to them, on a scale of 1 (totally unsatisfactory—"there is no way I would visit this place") to 10 (totally satisfactory—"I will definitely visit this place"). The recommendations offered to the users covered 90% of the taxonomy level-2 category of places (from bars, pizza places and museums, to casinos and zoos). The 40 recommendations assessed by each user were generated using five different algorithms, with each algorithm having generated 8 of the recommendations. The algorithms that were used to generate the recommendations are:

1. the proposed algorithm,
2. the proposed algorithm, modified to consider profile and place information at taxonomy level 2, instead of taxonomy level 3,
3. a plain CF algorithm (the algorithm in Sect. 2.4 taking only the cumulative influence into account and not considering the QoS dimension),
4. a plain QoS-based algorithm (the algorithm in Sect. 2.4 without the CF dimension) and
5. the proposed algorithm, but without considering per-category influencers for each user, but using a single set of influencers per user, across all categories.

Recommendations were presented to the users for assessment in randomized order.

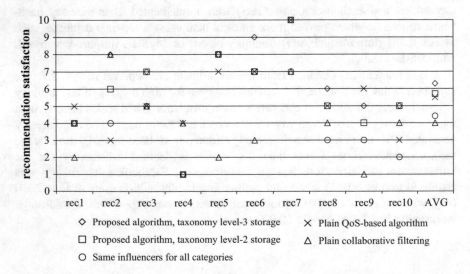

Fig. 2.8 Users' satisfaction of recommendations made by individual recommendation algorithms

Figure 2.8 depicts the participants' satisfaction regarding the recommendations they received, as measured in this experiment. On average (last column on Fig. 2.8) it is clear that the proposed algorithm using the taxonomy level-3 category storage outperforms the other algorithms, attaining an overall user satisfaction of 6.3. The proposed algorithm using the taxonomy level-2 comes in second with an overall user satisfaction of 5.7, or 90% of the satisfaction of the same algorithm with the taxonomy level-3 category storage. The plain QoS-based algorithm was ranked 3rd, the proposed algorithm modified to use a single set of influencers was ranked 4th, and the plain collaborative was ranked 5th, with their satisfaction being at the 87, 70 and 63%, respectively of the proposed algorithm with the taxonomy level-3 category storage. This experiment clearly shows that using a specialized set of recommenders for each place category provides a significant improvement in the quality of the generated recommendations.

Within Fig. 2.8 we have also included user ratings for 10 individual recommendations (rec1–rec10); these have been chosen to demonstrate that algorithm performance is not uniform across all cases. In our future work, we will further investigate cases where the proposed algorithm's recommendation received a poor rating (inferior to the ratings of other algorithms or lower than 5 out of 10).

2.6   Conclusions and Future Work

In this chapter we have presented a knowledge-based algorithm for generating leisure time place recommendation for SN, taking into account (a) qualitative characteristics, (b) the users visiting habits, (c) the influencing factors between social network users and (d) semantic information, concerning both the places to be recommended and the places' geographic locations. The recommendation algorithm follows the metasearch algorithm paradigm, using two different ranking algorithms, the first examining only the qualitative aspects of places and the related users' habits and the second being based on CF techniques. The rankings produced by these two algorithms are combined to generate the overall ranking, which is then used for generating the recommendations.

The proposed algorithm contributes to the state-of-the-art by considering qualitative aspects of the places, the influencing factors between social network users, the social network user past behavior regarding their visits in different place categories, and the semantic categorization of the places to be recommended. Furthermore, influencers in this algorithm are considered per category, to allow for formulation of more accurate recommendations and maximize the probability that recommended place are visited. The proposed algorithm has been experimentally validated regarding (i) its performance, and (ii) recommendation accuracy (users' satisfaction to the recommendations produced) and the results are encouraging.

Regarding our future work, we plan to conduct a user survey with a higher number of participants and more representative demographics. The current participants set was drawn from the University of Athens community, hence it is not a representative sample of the overall population; thus, the results drawn need further verification regarding their generalizability. A more comprehensive survey will tackle this issue and allow us to gain better insight into the satisfaction and needs of users with more diverse profiles. The planned survey will allow deeper analysis regarding the value of 0.68, which has been used in the experiments described in Sect. 2.5 as the QoS similarity threshold for places. This value has been derived by a small-scale experiment, in which participants were asked to rate whether 100 items were "close" or not to their QoS preferences, and then the QoS threshold was set to the value that maximized the QoS-predictions' F1-measure [40]. The QoS-prediction considered here is that a place p is considered "close" if its $score_{QoS,p,u}$ is greater than or equal to the QoS threshold and "not close" if its $score_{QoS,p,u}$ is less than the QoS threshold. The extended survey will allow us to obtain a more comprehensive dataset regarding the "closeness" perception and further investigate this issue. The main aspects of this investigation are (a) the value of the QoS threshold and (b) whether the QoS threshold is uniform across all categories and/or user profiles.

Furthermore, the QoS threshold mechanism leads to a behavior that the user is confined to viewing only information about places similar to those she has visited in the past, limiting thus the serendipity that may stem from recommendations, which is a desirable feature of RS [42]. To this end, means for allowing serendipity in recommendations, such as hybrid systems [23] or item-to-item mechanisms [15] will be examined. To alleviate the grey sheep problem, we will consider performing comparisons at higher taxonomy levels, when comparisons at the third level of the taxonomy lead to very small numbers of near neighbors.

We finally plan to take into account keywords of the descriptions ("hated that place", "I will never go again", "I loved that place", "best café ever", "perfect atmosphere", etc.) that follow check-ins and tags, as well as asking users to explicitly evaluate the QoS attributes of each place they visit, in order to achieve more accurate recommendations.

References

1. Burke, R.: Knowledge-based recommender systems. In: Kent, A. (ed.) The Encyclopedia of Library and Information Science. Marcel Decker Inc., U.S. (2000)
2. Blanco-Fernández, Y., Pazos-Arias, J.J., Gil-Solla, A., et al.: A flexible semantic inference methodology to reason about user preferences in knowledge-based recommender systems. Knowl.-Based Syst. **21**(4), 305–320 (2008)
3. Aggarwal, C.C.: Knowledge-based recommender systems. In: Recommender Systems. Springer, Berlin. ISBN: 978-3-319-29657-9
4. Facebook: Facebook ad targeting. https://www.facebook.com/business/products/ads/ad-targeting (2015)
5. He, J., Chu, W.W.: A social network-based recommender system (SNRS). Ann. Inform. Syst. **12**, 47–74 (2010)
6. Arazy, O., Kumar, N., Shapira, B.: Improving social recommender systems. IT professional, September (2009)

7. Oechslein, O., Hess. T.: The value of a recommendation: the role of social ties in social recommender systems. In: 47th Hawaii International Conference on System Science (2014)
8. Quijano-Sanchez, L., Recio-Garcia, J.A., Diaz-Agudo, B.: Group recommendation methods for social network environments. In: 3rd Workshop on Recommender Systems and the Social Web within the 5th ACM International Conference on Recommender Systems (RecSys'11) (2011)
9. Boulkrinat, S., Hadjali, A., Mokhtari, A.: Enhancing recommender systems prediction through qualitative preference relations. In: 11th International Symposium on Programming and Systems (ISPS), pp. 74–80 (2013)
10. Margaris, D., Georgiadis, P., Vassilakis, C.: A collaborative filtering algorithm with clustering for personalized web service selection in business processes. In: Proceedings of the IEEE 9th RCIS Conference, Athens, Greece (2015)
11. Bakshy, E., Rosenn, I., Marlow, C., Adamic L.: The role of social networks in information diffusion. In: Proceedings of the 21st International Conference on World Wide Web, pp. 519–528 (2012)
12. Bakshy, E., Eckles, D., Yan, R., Rosenn I.: Social influence in social advertising: evidence from field experiments. In: Proceedings of the 13th ACM Conference on Electronic Commerce (2012)
13. Schafer, J.B., Frankowski, D., Herlocker, J., Sen, S.: Collaborative filtering recommender systems. In: The Adaptive Web, LNCS vol. 4321, pp. 291–324 (2007)
14. Zhang, W., Chen, T., Wang, J., Yu, Y.: Optimizing top-n collaborative filtering via dynamic negative item sampling. In: Proceedings of the 36th International ACM SIGIR (SIGIR'13), pp. 785–788 (2013)
15. Herlocker, J.L., Konstan, J.A., Terveen, L.G., Riedl, J.T.: Evaluating collaborative filtering recommender systems. ACM TOIS 22(1), 5–53 (2004)
16. Balabanovic, M., Shoham, Y.: Fab: content-based, collaborative recommendation. Commun. ACM 40(3), 66–72 (1997)
17. Rodríguez-González, A., Torres-Niño, J., Jimenez-Domingo, E., Gomez-Berbis, M.J., Alor-Hernandez, G.: AKNOBAS: A knowledge-based segmentation recommender system based on intelligent data mining techniques. Comput. Sci. Inform. Syst. 9(2), (2012)
18. Monfil-Contreras, E.U., Alor-Hernández, G., Cortes-Robles, G., Rodriguez-Gonzalez, A., Gonzalez-Carrasco, I.: RESYGEN: a recommendation system generator using domain-based heuristics. Expert Syst. Appl. 40(1), 242–256 (2013)
19. Konstas, I., Stathopoulos, V., Jose, J.M.: On social networks and collaborative recommendation. In: Proceedings of the 32nd International ACM SIGIR Conference on Research and Development in Information Retrieval. Boston, USA (2009)
20. Jamali M., Ester, M.: A matrix factorization technique with trust propagation for recommendation in social networks. In: Proceedings of the fourth ACM Conference on Recommender Systems, RecSys 2010. Barcelona, Spain (2010)
21. Zheng, Y., Xie, X.: Learning travel recommendations from user-generated GPS traces. ACM Trans. Intell. Syst. Technol. (TIST) 2.1 (2011)
22. Bao J., Zheng Y., Mokbel M.: Location-based and preference-aware recommendation using sparse geo-social networking data. In: Proceedings of the 20th International Conferences on Advances in Geographic Information Systems, SIGSPATIAL'12, pp. 199–208 (2012)
23. Colombo-Mendoza, L.O., Valencia-García, R., Rodríguez-González, A., Alor-Hernández, C., Samper-Zapaterd, J.J.: RecomMetz: a context-aware knowledge-based mobile recommender system for movie showtimes Expert Syst. Appl. 42(3), 1202–1222 (2015)
24. Yang, W.-S., Hwang, S.-Y.: iTravel: a recommender system in mobile peer-to-peer environment. J. Syst. Softw. 86(1), 12–20 (2013)
25. Moreno, A., Valls, A., Isern, D., Marin, L., Borràs, J.: SigTur/E-destination: ontology-based personalized recommendation of tourism and leisure activities. Eng. Appl. Artif. Intell. 26(1), 633–651 (2013)
26. Ference, G., Mao, Y., Lee, W-C.: Location recommendation for out-of-town users in location-based social networks. In: Proceedings of ACM CIKM13, pp. 721–726 (2013)

27. Gilbert, E., Karahalios, K.: Predicting tie strength with social media. In: Proceedings of the SIGCHI Conference on Human Factors in Computing Systems (CHI'09), pp. 211–220 (2009)
28. Anagnostopoulos, A., Kumar, R., Mahdian, M.: Influence and correlation in social networks. In: Proceedings of the 14th ACM SIGKDD (KDD'08), pp. 7–15 (2008)
29. Facebook: Facebook interest targeting, https://www.facebook.com/help/188888021162119 (2015)
30. Chedrawy, Z., Abidi, S.S.R.: A web recommender system for recommending, predicting and personalizing music playlists. In: Proceedings of Web Information Systems Engineering (WISE 2009), pp. 335–342 (2009)
31. Aslam, J., Montague, M.: Models for metasearch. In: Croft, W.B., Harper, D.J., Kraft, D.H., Zobel, J. (eds.) Proceedings of the 24th Annual International ACM SIGIR 2001, pp. 276–284 (2001)
32. Pirasteh, P., Jung, J.J. Hwang, D.: Item-based collaborative filtering with attribute correlation: a case study on movie recommendation. In: 6th Asian Conference, ACIIDS 2014, Bangkok, Thailand, 7–9 April 2014, Proceedings, Part II, pp. 245–252 (2014)
33. Androutsos, D., Plataniotis, K.N., Venetsanopoulos, A.N.: Distance measures for color image retrieval. In: Proceedings of the International Conference on Image Processing, vol. 2, pp. 770–774 (1998)
34. Jones, C.B., Alani, H., Tudhope, D.: Geographical information retrieval with ontologies of place. In: Proceedings of the Conference on Spatial Information Theory, COSIT'01, pp. 322–335 (2001)
35. Word2Vec Library. https://code.google.com/archive/p/word2vec/ (2013)
36. ITU. Recommendation E.800 quality of service and dependability vocabulary (1988)
37. Mersha, T., Adlakha, V.: Attributes of service quality: the consumers' perspective. Int. J. Serv. Ind. Manage. 3(3), 34–45 (1992)
38. Margaris, D., Vassilakis, C., Georgiadis, P.: An integrated framework for adapting WS-BPEL scenario execution using QoS and collaborative filtering techniques. Sci. Comput. Program. 98, 707–734 (2015)
39. He, D., Wu, D.: Toward a robust data fusion for document retrieval. In: IEEE 4th International Conference on Natural Language Processing and Knowledge Engineering—NLP-KE (2008)
40. Lipton, Z.C., Elkan, C., Naryanaswamy, B.: Optimal thresholding of classifiers to maximize F1 measure. In: Proceedings of ECML PKDD 2014 (part II), pp. 225–239 (2014)
41. Data Center Knowledge: The Facebook data center FAQ. http://www.datacenterknowledge.com/the-facebook-data-center-faq/ (2013)
42. Ge, M., Delgado-Battenfeld, C., Jannach, D.: Beyond accuracy: evaluating recommender systems by coverage and serendipity. In: Proceedings of RecSys '10, pp. 257–260 (2010)

Chapter 3
An Ontology Based System for Knowledge Profile Management

A Case Study in the Electric Sector

Oscar M. Rodríguez-Elias, Maria de Jesús Velázquez-Mendoza and Cesar E. Rose-Gómez

Abstract Knowledge has become one of the most important factors for organizational success. In organizations, the most important knowledge remains in people: their employees and collaborators. Thereafter, for a knowledge management initiative to be successful, it should be focused on people. Hence, there exist an important relationship between knowledge management and human resource management. Getting answers to questions such as: what knowledge a specific job into the organization requires? What knowledge the people in charge of specific activities have? What knowledge a person needs to perform a better job? And so on, could be important for selecting the best candidates to achieve specific roles or positions into an organization. In this chapter, an ontology for knowledge profiles management for getting answers to such kind of questions is presented. The ontology was proved towards its implementation using real data obtained from a case study, which helped us to validate that it is actually useful to get answers to the kind of questions that guided its design.

Keywords Knowledge profile · Knowledge workers management · Knowledge profile ontology · Human resources management

O.M. Rodríguez-Elias (✉) · Maria de JesúsVelázquez-Mendoza · C.E. Rose-Gómez
División de Estudios de Posgrado e Investigación, Tecnológico Nacional
de México—InstitutoTecnológico de Hermosillo, Av. Tecnológico S/N,
Hermosillo C.P. 83170, Sonora, Mexico
e-mail: omrodriguez@ith.mx

Maria de JesúsVelázquez-Mendoza
e-mail: rvelazqu@ith.mx

C.E. Rose-Gómez
e-mail: crose@ith.mx

© Springer International Publishing AG 2017   49
G. Alor-Hernández and R. Valencia-García (eds.), *Current Trends
on Knowledge-Based Systems*, Intelligent Systems Reference Library 120,
DOI 10.1007/978-3-319-51905-0_3

3.1 Introduction

Human resource management is a vital process to determine the accomplishment of the goals that guide the maintenance and development of a successful organization, since human resources are an essential and strategic part of every organization [1]. Some key processes of human resources are comprehension, creation of new ideas, problem solving, decision-making, etc. [2].

The relevance of human resources tends to increases every day because of the knowledge they possess, which is becoming too valuable for organizations. This has conducted to see human resources as knowledge workers or talents (see for instance [3, 4]). Therefore, for an organization to better manage their knowledge workers, a mean for describing the knowledge profile of their employees would be necessary. The organization would need to know: what knowledge their employees possess?, what knowledge they require to accomplish the tasks they are in charge of in a better way?, how to capacitate better their employees in order for they to acquire the knowledge they require for becoming more productive, or scaling into the enterprise?, and so on.

There exist different approaches for helping organizations to identify the above information, such as Knowledge Audit Methodologies [5]. Nevertheless, although some of them provide technological means for conducting such type of studies [6], in literature we have not found one that specifically aid organizations to manage the knowledge profiles of their employees.

The last observation got us to conduct a research to develop a knowledge-profile management system based on a model for knowledge profile definition [7]. The result of this effort was the development of an ontology for describing knowledge profiles and a prototype for a knowledge-based system based on such ontology. The ontology and the prototype were used to analyze the knowledge profiles of the employees of an electricity generation organization. In this paper, we describe the development and implementation of the ontology, and the main results of its application in a case study.

This paper is organized as follows: first, we present the theoretical background that justifies our proposal, followed with a state of the art section, where we aboard related work on the use of ontologies for knowledge management, and for people profile definition. Later, we detailed describe the development of the ontology. After that, we introduce the main results of our research, in which we illustrate the implementation of the ontology into a knowledge-based system, using information form a real case. Finally, our conclusions are presented.

3.2 Theoretical Background

Tacit knowledge, the one that remains in people, is commonly the most valuable knowledge for organizations, because it allows them to innovate, and facilitates fast adaptation to changes, however, because its nature, it is difficult to manage [8]. The last joined to the growing importance of knowledge for organizations, has guided changes to the way human resources are managed, causing employers to see human resources as knowledge workers, not only as workforce [9].

Human resource management integrates several activities and processes, between the ones we can mention [10, 11]:

- Activities related to the job. The selection of people with the appropriate skills, knowledge and expertise to fill job vacancies. Some practices included in this activity are human resource planning, jobs analysis, and personnel recruitment and selection.
- Human resource development. Including analysis of training or capacitating needs, to assure that employees have the needed knowledge and skills to fulfill their jobs in a satisfactory fashion; this involve identifying the key skills and competencies to develop.
- Employees' remuneration or retribution policies. This involves planning what remuneration or benefits to offer to employees.
- Others related to employees satisfaction, such as labor-management relations, or health and safety.

It can be seen that a big part of the activities of the human resource management discipline are related to the identification and definition of knowledge, skills, aptitudes, etc. needed to accomplish the required activities for a job in a satisfactory way, or to identify the best candidate for a specific job. Additionally, it is also important to identify personnel's weaknesses or deficiencies, in order to provide the corresponding training or capacitation, making people more able to better perform its work activities, or to aspire to better jobs into the organization.

Peter Drucker stated that the major contribution in organizations' management in the XXI century, should be increasing productivity of knowledge work and knowledge workers [12]. Start seeing human resources as knowledge workers, not only as workforce, has important implications to the way human resources are managed in organizations. Traditionally, human resources management looks for employees capable of following orders, establishing what people must do during their working time, and how that work should be done. However, as work become more complex, it also become more knowledge intensive [13], requiring people with skills, abilities, aptitudes, attitudes, capabilities, etc. needed to make every time more difficult decisions about how to do their work. They must decide when, where, why and under what conditions to do their working activities, not only know what to do [12]. Therefore, it is not only needed to know the academic credentials and career path of individuals, but also to know their skills, competencies, formal and informal knowledge, values, expertise, culture, etc., in order to be able to identify the people that better fit to the needs of working positions and roles [14]. This makes human resource management a more difficult task, since it makes more difficult to identify the best people for a job, or to maintain the best workers into the organization [15].

The human resource management field has evolved looking solutions to the kind of problems described. For instance, Lengnick-Hall et al. have observed a change in the way human resources are seen in strategic human resource management field. These authors state that initially, the field of human resources was focused on providing employees with skills and motivation needed to achieve established organizational goals, and on assuring existence of enough employees with specific knowledge required to solve organizational necessities. However, strategic human resource management has changed its focus to one centered on the contributions of human capital, strategic capabilities, and organizational competitive performance [16].

Talent management is another result of the efforts to see human resource from a different perspective; one centered on the potential of human resources to contribute in a strategic manner to productivity and organizational growing [17]. Talent refers to all those employees that add critical value to every place in the organizational structure, making actions in planed and predictable periods. According to [18], organizations have big difficulties when they do not have a clear and operative definition about talent management.

All these challenges lead to the development of novel tools capable of integrating automatic structures to support functional strategies of organizations, and that allow conceptualize talent integrating it as part of the employees' profiles. Such profiles not only should consider the academic formation of employees, but also skills, attitudes, capabilities, competencies, formal and informal knowledge, values, between other important elements; what we have defined as knowledge profile. We consider a knowledge profile as "a set of structured features that describe the needed knowledge, related to resources and capacities, that allow to habilitate dynamic generation of work competencies, competencies for key processes, and distinctive competencies that add value to organizations" [19].

Therefore, a knowledge profile not only should consider the knowledge required for a job or the one a person dominates, but also her competencies, skills, values, etc. Hence, defining a knowledge profile for a role or job, as well as determining if a person could fill such profile, becomes a complex task. The last is because of the wide diversity of knowledge subjects, skills, competencies, values, etc. This becomes even more complex if we consider that such knowledge, skills etc. could be present at different levels on each individual, as well, each role or job could require people with different levels of domination for each knowledge feature. In order to better manage such complexity, we consider that it is required a common language and structure that facilitate the design of technological systems capable of supporting the management of knowledge profiles in organizations.

Artificial intelligence provides tools and techniques useful for supporting several problems faced in the human resource management field [20]. In our context, we have observed that the use of an ontology could be a good solution to the problem of defining knowledge profile in a structured manner.

3.3   State of the Art

Ontology has different meanings according to the community in which the concept is defined. However, in a general point of view, an ontology is used to formally describe the "nature and structure of things" in terms of categories and relationships. An ontology is a mean for providing a common understanding through a shared conceptualization expressed in a formal and explicit specification [21]. In the computing field, perhaps the most used definition is the one of Gruber [22]: "an ontology is an explicit specification of a conceptualization". In the computing field, an ontology is useful to formally model the structure of a system or its information or domain knowledge [21], so they become an important mean for knowledge sharing [22]. Therefore, they are often used as a basis for developing knowledge-based systems, and more recently for knowledge management systems, as it is the case of our approach.

3.3.1   Ontologies for Knowledge Management

Ontologies have become one of the most used means for representing, organizing and structuring knowledge in several domains [23], so they are an important tool to aid in the design of knowledge management initiatives. Some of the efforts are made to apply ontologies to facilitate interoperation of heterogeneous information systems, and in his way, be able to share knowledge within an entire organization, or between different organizations. One of the ways for doing the last is by integrating ontologies and multi agent systems as shown in [24]. The usefulness of integrating ontologies in multi-agent systems for knowledge management purposes has been demonstrated previously in [25], with the development of a prototype to manage knowledge in software maintenance process. Because of its knowledge intensive nature, software development and maintenance is a common field for developing ontologies for managing software process knowledge [26, 27].

Another research field on the use of ontologies for knowledge management is by integrating ontologies and web technologies [28]. Within this context, in [29] an ontology to manage data representation and the structure of supply chain knowledge is used as a part of an approach for applying semantic web technologies for knowledge management in supply chains. Semantic web technologies are one of the main application areas of ontologies, since they provide means to improve searching information over the Internet. As noted in [30] "Semantic Web Technologies allow combine architectural styles for software development such as Service-Oriented Architecture (SOA) and Event-Driven Architecture (EDA)", aiding in the development of several solutions for web based business services.

The evolution of Internet services has driven to a new model of web services called Cloud computing, which refers to a new model of computing services providing, where software, platform or computing infrastructure is provided as a service consumed on demand over the Internet [31]. In [32] ontologies are used as a mean to provide a web semantic platform to assist in the process of discovering the cloud services that best match user needs, by an automatic semantic annotation platform that can assists users in the process of adding metadata to web content including cloud services descriptions. These authors also propose a semantic search engine to improve the precision and recall of the cloud services search results by leveraging the cloud service-related semantic annotations. In [33] these same authors show the importance on integrating ontology evolution to their approach, since this allows an ontology to be timely adapted, and the changes to dependent artifacts be consistently propagated.

Semantic web technologies, together with the fast growing of information all over the Internet, provide powerful means for leveraging global knowledge. A good example is [34], where ontologies and semantic web technologies are used to develop a system to facilitate access to Wikipedia information by means of natural language capabilities. As well, in [35] ontologies and web semantic technologies are used to improve retrieval of engineering domain knowledge.

From a different field, but related to semantic technologies for business process improvement, in [36] ontologies are proposed as a mean for modeling business processes, easing semantic ambiguity in business process models. The last is accomplished since ontologies allow including semantic information into the process models. In the context of business process modeling and improvement, one of the most related works to ours' is presented in [6], which consists of an ontology for representing the results of a knowledge audit process, considering concepts such as process, knowledge involved into the process, and agents having the knowledge. This ontology is validated through its implementation as a prototype in the protégé knowledge base system. Although this ontology includes a class to identify people with important knowledge for organizational processes, it does not provide means for representing the knowledge profile of such people.

As can be observed in this subsection, ontologies have several applications to improve business processes, however, as noted in [37], many efforts to use ontologies in business have been less successful than expected; and on the other hand, the main benefits of using ontologies are defined and categorized as technical-centered or user-centered. Therefore, we have decided to consider this observation, and have focused on the development of a specific domain and a technical centered ontology for knowledge profile management.

3.3.2   Ontologies for Users' Profile Management

Because of the focus of our proposed ontology, another related field is the use of ontologies for representing people profiles. In this field, we have found several researches that principally aboard the management of users' profiles, not knowledge profiles, either human resources profiles. In [38] ontologies are used to generate users' profile based on the users' interest in order to improve information retrieval systems. Authors propose to construct the user profiles from web resources, such as personal web pages related to their name, title, research interests, publications, research projects, and so on. The user profile is defined from several user ontologies, and each specific ontology can be shared by several user profiles. In [39] a different approach for generating users' profile is presented; these authors proposes the use of knowledge extraction techniques applied to a general ontology, in order to construct a personal ontology for each user, defining in that way, a structured and semantically coherent representation of the user interests. In [40] a user profile ontology is also proposed and used to represent interactions between the users and their context in collaborative learning environments; this user profile ontology includes personal information; demographic information such as age, birthplace; preferences, interests, skills, current position, and academic degree.

Ontologies for describing people profiles are commonly focused on profiling users of information retrieval systems, in order to improve the search of information according to the users' interests. Therefore, such ontologies are not well sited for describing the knowledge profile of employees of an organization. Knowledge profiles should include descriptions of the skills, competencies, values, capabilities and other important strategic aspects for organizational success. Although we have found efforts to develop ontologies for describing such kind of topics, these works are centered on profiling the capabilities and competencies of organizations for supporting strategic management initiatives [41], we have not found a work focused on using ontologies for employees' knowledge profile management.

3.4 An Ontology for Knowledge Profile Management

The presented work was conducted using an Action Research approach [42]. It was a combination of theoretical analysis and practical application, as shown in Fig. 3.1. In general terms, the process followed can be divided in two main phases: first, we defined the ontology for knowledge profiles description following the stages proposed by Methontology [43]. On the other hand, we developed a prototype of a knowledge-based system using Protégé 4.1.0 (http://protege.stanford.edu/), to evaluate whether the ontology is useful to answer the type of questions considered in its specification.

To perform the evaluation of the ontology and the prototype, we applied them both in a case study conducted in an electricity generation enterprise to analyze the knowledge profiles of the people in charge of the electricity generation process. To obtain the information and conduct the case study we followed an adaptation of the KoFI methodology proposed in [44].

The remains of this section describe in detail how the ontology was developed, and its implementation in protégé.

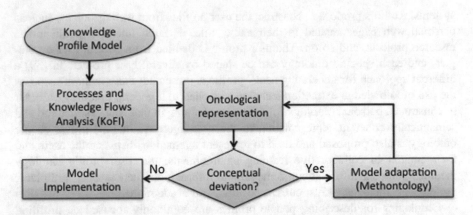

Fig. 3.1 General view of the process followed to develop the ontology

3.4.1 Development of the Ontology

Methontology [43] considers an iterative life cycle, that helps add, change, or eliminate terms on each version. Following this methodology, we describe the phases of specification, conceptualization, and implementation.

3.4.1.1 Specification

To identify the elements to consider for the ontology, we defined a set of questions related to the definition of the knowledge profiles for an employee or a candidate, and for the roles that those employees should fulfill. Table 3.1 shows the questions used.

3.4.1.2 Conceptualization

This activity consisted on organizing and converting the informal perception of a domain into a semiformal specification. This was done by means of a set of intermediate representations (tables and diagrams) easy to understand, either by domain experts or ontology developers. For this activity, Methontology provides a set of 11 tasks. Since we decided to develop a generic ontology, we did not realized tasks 8 and 11. The 10 tasks we realized are described as follows:

Task 1: Glossary of terms. The basic components or vocabulary of the ontology was defined, getting a glossary of 25 generic terms related to a knowledge profile definition. An extract of this glossary is shown in Table 3.2.

Task 2: Taxonomy of concepts. The taxonomy of concepts generated consists of the hierarchy of concepts included in the glossary of terms. To construct this

Table 3.1 Questions used to define the scope of the ontology

Related to the person (x)	Related to the role (y)	Related to the person (x) and the role (y)
What training has the person (x)?	What training is required for the role (y)?	What person(x) could fill the role(y)?
What knowledge has the person (x)? At what level?	What knowledge is required for the role (y)? At what level?	What roles (y) could be filled by the person (x)?
What resources dominates the person (x)? At what level?	What resources uses the role (y)? At what level?	What knowledge is missing on the person (x) to fill the role (y)?
What tasks performs the person (x)? At what level?	What tasks performs the role (y)? At what level?	

Table 3.2 Small extract of the glossary of terms used to construct the ontology

Name	Synonyms	Acronyms	Description	Type
Resource			Elements interacting into the functional part of an organization	Concept
Document			Documented information sources to aid in providing key knowledge, for instance, manuals, procedures, norms, etc.	Concept
Person	Individual, employee, staff, human resource	HR	A human being with individual capacities and in constant development that contribute to the organization's grow	Concept

hierarchy, we followed the taxonomic relationships defined in the Frame ontology of [45], and in the OKBC Ontology (OKBC = Open Knowledge-Base Connectivity) [46]: Subclass-of, Disjoint-Decomposition, Exhaustive-Decomposition, and Partition. Figure 3.2 shows a part of the hierarchy of concepts, defined using the whole glossary, whose extract was shown in Table 3.2.

Task 3: Ad hoc binary relationships diagram. The goal of this diagram is to establish the ad hoc relationships that exist within the same or different taxonomies of concepts. Figure 3.3 shows this diagram developed using CmapTools [47].

This schema illustrates the implicit link between the knowledge acquired by an individual and that required for a role. From the point of view of the individual, a person *has* studies (educational level, academic grade) and has been *trained-in* (knowledge area and discipline), required characteristics for a knowledge profile that is suitable for a role.

From the point of view of the organization, a job is composed of several roles; each role performs some tasks into a process. To complete those tasks some resources are required (documents, tools, other people), as well as some knowledge,

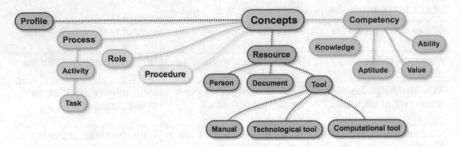

Fig. 3.2 Extract of the concepts' hierarchy

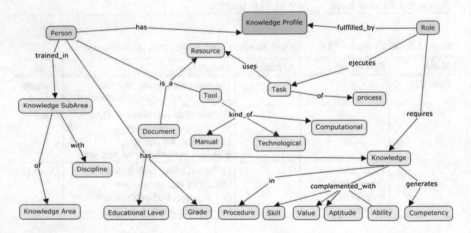

Fig. 3.3 Ad-hoc relationships diagram

which is complemented with skills, values and aptitudes, that allow attaining competencies.

Using the ad hoc relationships diagram, we detailed each binary relationship that aid associating the knowledge acquired by an individual, and that which is required for a role, as shown in Fig. 3.4a. Additionally, there are the implicit association of the knowledge level that allows describing the profile of resources and capacities, under the perspectives of the organization or the human resources, see Fig. 3.4b. On one side, a person has knowledge (complemented with skills, values, aptitudes, and so on) related to certain procedures at a level that generates competencies. On the other hand, for an organization, one could have a role that requires knowledge in certain procedures to execute tasks at a desired level.

Task 4: Dictionary of concepts. Using as a basis the concepts taxonomy and the relationships diagrams, we proceeded to specify the properties that describe each concept of the taxonomy, and the relationships identified. The Table 3.3 details the concepts of the domain, their instances, attributes of class and instance, and the relationships in which the concept is the origin of them.

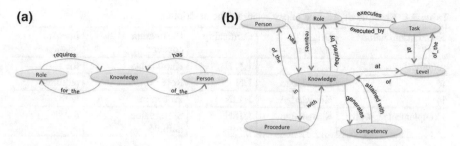

Fig. 3.4 Binary relationships diagrams: **a** for role-knowledge, knowledge-person, **b** for the knowledge level required for a role, or acquired by a person

Table 3.3 Example of domain concepts description

Concept name	Instance	Class attribute	Instance attribute	Relationships
Knowledge			Knowledge_Id	at, Complemented_with, of, in, generates, to
Discipline			Discipline_Id	of
Document		Resource_Type	Document_Id	is_a
Grade	Technician, bachelor, master, doctorate		Grade_Id	of
Skill			Skill_Id	to
Tool		Resource_Type		kind_of, is_a

Task 5: Ad_hoc binary relationships description. To formalize the captured knowledge, we also realized a detailed description of the ontology's ad_hoc binary relationships. We specified each relationship following the template shown in Table 3.4, considering the name of the relationship, the origin concept, the destination concept, the cardinality, and the inverse relation.

To complement the relations that allow defining the knowledge level required, we also included the Table 3.5. We used it to specify the concepts relationships as shown in Table 3.6, where it is described the normal relationship "Person has Knowledge at Level" and its inverse relationship "Level of Knowledge of Person".

The sets of relationships identified are those used to associate the knowledge acquired by an individual or required for a role, and to associate the control (mastery, domination level) of resources. For instance, to identify the domination level required by an individual for using resources during the execution of a task.

Task 6: Instance attributes. In this task, the instance attributes included in the concepts dictionary are detailed. The concepts instances and their values are

Table 3.4 Example of Ad_hoc binary relationships description

Relationship name	Origin concept	Cardinality	Destination concept	Inverse relation
requires	Role	1:1..N	Knowledge	for
to	Knowledge	1:N	Role	requires
in	Knowledge	1:1..N	Procedure	with
Complemented_with	Knowledge	1:1..N	Skill, value, attitude	for

Table 3.5 Description of relationships for knowledge level

Name of addition relation	Destination concept	Cardinality	Addition concept	Inverse addition relation
at	Knowledge	1:1	Level	of
at	Task	1:1	Level	of

Table 3.6 Example of detailed description of normal and inverse relationships

Origin concept	Person
Relation	has
Inverse relation	of
Destination concept	Knowledge
Addition relation	at
Inverse addition relation	of
Addition concept	Level

Table 3.7 Example of instance attributes description

Attribute name	Concept	Value type	Cardinality
Aptitude_id	Aptitude	String	(1:1)
KnowledgeArea_ID	Knowledge area	String	(1:1)
Competency_Id	Competency	String	(1:1)
Knowledge_Id	Knowledge	String	(1:1)

described. Table 3.7 illustrates how this was done; it shows, as unique entities, some of the instances for the ontology.

Task 7: Class attributes. Attributes that represent generic characteristics of a concept, hence, they take values within the class where they are defined. All instances of a concept maintain the same value for these attributes. Class attributes are not inherited by the subclasses or instances. We described the class attributes included in the concepts dictionary. Table 3.8 illustrates how this was done.

Task 9: Formal axioms' description. Axioms are logic expressions that are always true. They are commonly used to specify the ontology's restrictions. During this task, we specified formal axioms from the point of view of the role, the

Table 3.8 Example of description of class attributes

Attribute name	Concept	Type of value	Cardinality	Values
Resource_Type	Person	[Person.Document.Tool]	[1..1]	Person
Resource_Type	Document	[Person.Document.Tool]	[1..1]	Document
Resource_Type	Tool	[Person.Document.Tool]	[1..1]	Tool

Table 3.9 Axiom for the capacities of an individual

Axiom	Capacities of an individual (person)
Description	Elements that define the topics of knowledge with their complements and that are fulfilled by an individual
Logical expression	exist (?P,?K), (?P,?S), (?P,?A), (?P,?V) where ((Person (?P) and Knowledge (?K) and Procedure (?PR) and Skill (?S) and Aptitude (?A) and Value (?V) and has (?P, ?K) and in (?K, ?PR) and complemented_with (?K,?S) and complemented_with (?K,?A) and complemented_with (?K,?V) and of_the (?S,?P) and of_the (?A,?P) and of_the (?V,?P))
Concept variable	Person-P, Knowledge-K, Procedure-PR, Skill-S, Attitude-A, Value-V
Relationships	has, in, complemented_with, of_the

individual, and the knowledge level required in the association of both entities. Table 3.9 shows an example. It indicates the existence of elements that define the capacities and complements of an individual. This is done through the relationships: "Person has Knowledge in Procedure", "Knowledge complemented with Skills, Attitude, and Values, and fulfilled by a Person".

In this task we identified and described twelve axioms: (1) training for a role, (2) capacities required for a role, (3) resources dominated (controlled) by a role, (4) person's training, (5) person's capacities, (6) resources dominated (controlled) by a person, (7) level of knowledge required for a role, (8) level of knowledge of a person, (9) level of performance for a task required for a role, (10) level of performance attained by a person to perform a task, (11) resource's domination level required for a role, and (12) resource's domination level attained by a person.

Task 10: Rules description. In this task we identified the logical rules required for processing the information described using the ontology. These rules were described following a "if <condition> then <consequence>" format. The left part of the rule is a set of simple conditions, while the right part correspond to values of the ontology. Table 3.10 illustrates the above. It describes a rule that allows infer the person that has the knowledge required for a role. The conditional application of the relationships "Knowledge of the Person", "Knowledge required by a Role" and "Role fulfilled by a Profile", helps determine which Person can fill the Role.

In order to be able to give answers to the questions that relate the person and the role in Table 3.1, we defined three rules: (1) persons that fill the requirements for a

Table 3.10 Rule for inferring which person can fill a role

Rule	Person that fill the requirements for a role
Description	A profile of a role for which the required knowledge must be identified in a set of individuals

Expression	If (Role (?R) and Person (?P) and Knowledge (?K) and
	      Skill (?S) and Value (?V) and Aptitude (?A) and
	      Task (?T) and Resource (?RE) and
	        of (?L,?K) and required_by (?K,?R) and
	        complemented_with (?K, ?S) and
	        complemented_with (?K, ?V) and
	        complemented_with (?K, ?A) and
	        of_mastery_for (?L, ?T) and executed_by (?T, ?R) and
	        of_mastery_for (?L, ?RE) and for (?RE, ?R) and
	        has (?P, ?K) and at (?K, ?L) and
	        complemented_with (?P, ?S) and
	        complemented_with (?P, ?V) and
	        complemented_with (?P, ?A) and
	        executes (?P, ?T) and at (?T, ?L) and
	        control (?P, ?RE) and at (?RE, ?L))
	Then
	      [Person] (?P)

Concept-variable	Person-P, Role-R, Knowledge-K, Skill-S, Value-V, Aptitude-A, Task-T, Resource-RE, Level-L
Relationships	of, required_by, complemented_with, executed_by, for, has, executes, at, of_mastery_for, control

role, (2) roles that can be filled by a person, and (3) knowledge missing on a person to fill a role.

In the next section, as results of our work we present the implementation of the ontology used for conducting the case study, complemented with the main results obtained from performing such work.

3.5 Results

We consider that the main results of our study are the ontology validated through its implementation in a real case, and the result of the analysis of the jobs identified into the process. With this implementation, it was possible to evaluate that the ontology is useful to obtain answers to the questions proposed as a basis for its design. The implementation of the ontology consists of a protégé (version 4.1.0) [48] based prototype. It is important to mention that the application of the ontology to the case study was made in Spanish, so that the concepts and relationships implemented into the prototype are in their original language. We decided not to translate it in order to maintain the original data.

3.5.1   The Case Study

A case study was conducted to obtain real data to evaluate the feasibility of implementation of the ontology, and to validate that it actually helps to answer the kind of questions used as a basis for its design (see Table 3.1).

The case study was performed in an Electricity Generation Enterprise, and consisted on the analysis of the activities of the roles carried out by the people in charge of the Electricity Generation Process.

To obtain the data and perform the analysis of the process, we followed an adaptation of the KoFI methodology [44] for studying knowledge flows in organizational process. This analysis consisted on the study of the processes map of the organization, from which we identified the main sources of knowledge, the people in charge, the main activities, the main knowledge required for each task, and the main knowledge that the people in charge have. We used the Business Process Modeling Notation (BPMN) [49] as a tool for documenting the processes during the analysis.

From the analysis we identified six main jobs or positions in the process: Sub Manager of Operations (Job 1), Operations support (Job 2), Shift manager (Job 3), Operations Technician (Job 4), Operator (Job 5), and Results engineer (Job 6). As well, we found that all these six jobs perform the same three roles at different levels of expertise. The three roles they perform are Operator, Analyst, and Trainer. We used a scale of 4 levels of expertise for performing such roles, (1) elemental, (2) medium, (3) advanced and 4) expert, which are the ones used by the studied enterprise. Figure 3.5 shows the level of expertise by role required for each job.

Together with the roles carried out, we identified the main knowledge areas required for each job, the six most important knowledge areas, and the level required for each job are shown in Fig. 3.6. Other knowledge area identified were operative chemistry, fuel management, fluids and pneumatic, electricity and magnetism, engines and electricity generators, electric protections, instrumentation and control, electric net operation, environment, and communication techniques.

Fig. 3.5 Level of expertise by role required for each job

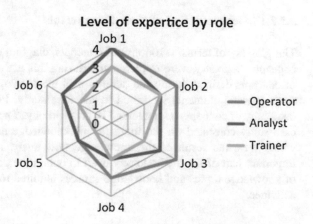

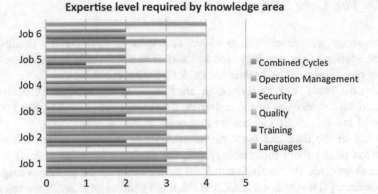

Fig. 3.6 Level of expertise by knowledge area required for each job

The information obtained from the case study was used to implement the ontology and tests if it is actually useful to infer answers to the questions that guided its development. Next we describe how this implementation was done.

3.5.2 Ontology Implementation

The ontology implementation consisted on the realization of the conceptualization phase of Methontology [50], that means: the determination of the base components, detailed definition of the base components, and normalization. We opted for using the Protégé support for OWL-DL (Ontology Web Language-Descriptive Logic) [51], because of its descriptive logic based on first order logic. The above facilitates the use of automatic reasoning for verifying the existence of inconsistencies in the concepts classification hierarchy.

3.5.2.1 Determination of Base Components

The glossary of terms, taxonomy of concepts, diagram of binary relationships, and concepts dictionary were implemented during this activity. The generic taxonomy of concepts defined during the definition of the ontology was extended to include some specific elements identified in the case study. Figure 3.7 shows the whole taxonomy of concepts implemented in the ontology. The extension made during the case study consisted on the inclusion of knowledge and resources classification, which were the result of the knowledge flow analysis made to the process. An important part of the methodology used to obtain the data [44] is the identification of knowledge topics and knowledge sources required to perform the main process' activities.

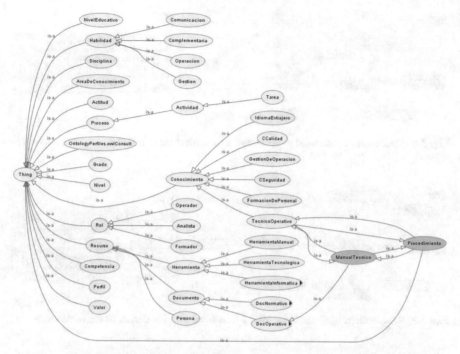

Fig. 3.7 Knowledge profile ontology for the operation of electricity generation process

Another inclusion to the ontology was the association of the "ManualTecnico" (Technical Manual) subclass to the classes "Procedimiento" (Procedure), "Conocimiento" (Knowledge), and "Documento" (Document). This was because the operative technical knowledge is documented in such manuals.

3.5.2.2   Detailed Definition of Base Components

The details of the ad hoc binary relationships were registered in the ontology implementation during this stage. It was necessary to rename these binary relationships because of some Protégé restrictions. Nevertheless, improved readability was obtained.

Since the organization does not have an identification of the level associated to knowledge or resources, the relationships related to that were not considered in the implementation. Nonetheless, we classified the knowledge subjects to be managed according to the level of knowledge associated to the roles.

To get a graphical view of the relationships of the ontology definition, we used the Ontograf Plug-in (version 1..1) of Protégé. Figure 3.8a illustrates this with the relationships that delimits the resources dominated by a person, the resources used by a role, and the tasks performed by a person and that are part of the activities of a

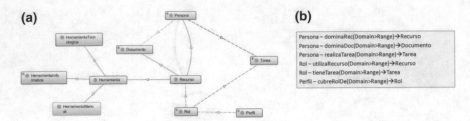

Fig. 3.8 Relationships associated to resources. **a** Graphical view. **b** Details of the relatiohsips

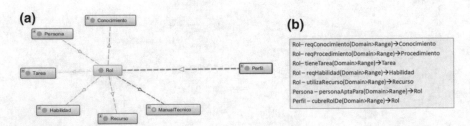

Fig. 3.9 Relationships associated to a role. **a** Graphical view. **b** Details of the relatiohsips

role. Figure 3.8b shows the details of these relationships. Complementarily, Fig. 3.9a illustrates the relationships of the role while Fig. 3.9b presents the details of such relationships. These relationships allow describing requirements for a role (knowledge, skills, resources to use, task to carry on, etc.) and to specify a person who fills a role (**Persona-personaAptaPara(Domain>Range)→Rol**).

Additionally, the relationships associated to a Person were also described, considering the elements to define the knowledge profile (knowledge, skills, values, aptitudes, procedures), resources dominated (Documents, Resources), and training (Educational Level, Knowledge Area, Discipline).

3.5.2.3 Normalization

The axioms used to delimit the rules of the ontology were registered in this stage. To accomplish this it was necessary to include additional classes. One of the benefits of using OWL-DL was that at the moment a reasoner is activated, a classes hierarchy is automatically generated, respecting the set of logic conditions that use such classes [52]. Table 3.11 shows the axioms that were implemented into the ontology.

To determine the consistency of classes and its relationships, and to infer knowledge from the registered specifications of the axioms, it is required to use a reasoner. In our case, the reasoner we used was Hermit OWL (version 1.3.6) [53].

Table 3.11 Axioms implemented in the ontology	*Axiom 2.* Capacities of a role (ActReqPor value Operador_elemental) or (ConocReqPor value Operador_elemental) or (HabReqPor value Operador_elemental) or (ProcedReqPor value Operador_elemental) or (ValorReqPor value Operador_elemental)
	Axiom 5. Capacities of an individual (ActitudDe value PR003) or (ConocDe value PR003) or (HabilidadDe value PR003) or (ProcedConocPor value PR003) or (ValorDe value PR003)
	Axiom 3. Resources required by a role (DoctoReqPor value Operador_elemental) or (RecReqPor value Operador_elemental)
	Axiom 6. Resources required by an individual (DocDominadoPor value PR003) or (RecDominadoPor value PR003)

Table 3.12 Rule to identify a person that fulfills the requirements for a role	*Rule 1.* Person that fulfil the requirements for a role
	((ActReqPor value Operador_elemental) and (ActitudDe value PR003)) or ((ConocDe value PR003) and (ConocReqPor value Operador_elemental)) or ((DocDominadoPor value PR003) and (DoctoReqPor value Operador_elemental)) or ((HabReqPor value Operador_elemental) and (HabilidadDe value PR003)) or ((ProcedConocPor value PR003) and (ProcedReqPor value Operador_elemental)) or ((RecDominadoPor value PR003) and (RecReqPor value Operador_elemental))

Using the reasoner on the axioms 2 and 3 it is possible to obtain the capacities and the level of domination of resources that a person has. With the axioms 5 and 6 one can identify the capacities and resources required for a role. The rule 1 shown in Table 3.12, illustrates the union of the axioms. It can be noted the use of specific instances in this rule, particularly the value PR003 is a reference to a specific person. It is necessary to directly indicate the instances into the expression, because DL Query, the standard query language of Protégé 4.1, uses the information related to the classes: their properties or instances.

Finally, each of the inferences defined were evaluated in order to identify if the ontology was correct for getting answers to the initial questions (see Table 3.1). As example, Fig. 3.10 shows the details of the axioms that compound the rule 1. Figure 3.10c represents the search for a person that could fill a role requiring knowledge of "Sistema de Aseguramiento de la Calidad" (Quality Assurance System). It can be seen that the inverse relationships "tieneConocimiento" (has Knowledge) and "requiereConocimiento" (requires Knowledge) (Fig. 3.10a)

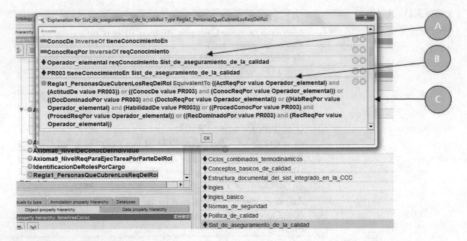

Fig. 3.10 Details of the axioms that compound the rule 1 executed in protégé

guided to the fact that such knowledge is required by the role "Operador Elemental" and that the person identified as "PR003" has such knowledge (Fig. 3.10b).

3.6 Discussion and Conclusion

In this paper we have presented an ontology developed to help manage knowledge profiles in organizations. The ontology was implemented by using information obtained from a real case, where the roles and knowledge required for the people in charge of the processes of an electricity generation enterprise were analyzed.

As was stated, we followed a set of questions for developing the ontology. The case study carried out helped us to validate that the ontology can actually help to obtain answers for such questions, for instance, which person has the knowledge required for a specific role? What roles a specific person can fill? What knowledge is required for a person to fill a specific role? And so on.

We consider that a tool such as the one we are proposing in this paper could have important applications in the field of knowledge management and human resource management. For instance, it could be useful to improve employee selection, assignation of personnel to work roles or positions, planning of employee career development, definition of training strategies or programs, among other uses.

Nevertheless, the case study also showed us an important limitation of our approach: it is important for a knowledge-based system using our ontology to solve real problems, to be able to support imprecision, ambiguity, or incomplete information. This is because in practice people do not fill the requirements of a role exactly at 100%. For instance, the level of knowledge that a person has could be lower or higher than the level required for the role.

Taking into account the last observation, we carried out a research work for developing a fuzzy logic model to compare knowledge profiles [54]. As future work, we want to integrate the fuzzy logic model into the ontology in order to be able to use it for creating better knowledge-based system to support human resource problems, considering the imprecisions found in practice.

Additionally, as part of our ongoing and further work, we are developing a web-based system that uses our ontology as a knowledge base for knowledge profile management. We are designing that system as a Cloud service, in order to use it as a basis for the development of several applications in the field of human resource management. Particularly, we want to explore the use of text mining and knowledge extraction techniques to automatically generate knowledge profiles from unstructured information contained in documents and web pages, for instance a curriculum vitae or a Linked-In profile, by means of the use of semantic web technologies. As well, and for evaluating the usefulness of the ontology for structuring curricular data in several domains, we are using it to compare the profiles of students of a university to those required by the organizations that usually hire them. With this last effort, we believe that our ontology could be a very useful mean to organize and structure the knowledge, hard and soft skills required in specific domains, in order to better coordinate the knowledge needs of the industry and the formation and capacitation proposals of universities.

Acknowledgements This work has been partially supported by Tecnológico Nacional de México, under grant 5335.14.15-PR.

References

1. Armstrong, M.: Strategic Human Resource Management: A Guide to Action, 3rd edn. Kogan Page, London (2006)
2. Dalkir, K.: Knowledge Management in Theory and Practice, 2nd edn. The MIT Press (2011)
3. Wiig, K.: People-Focused Knowledge Management: How Effective Decision Making Leads to Corporate Soccess. Elsevier, Amsterdam (2004)
4. Davenport, T.H.: Thinking for a Living: How to Get Better Performances And Results from Knowledge Workers. Harvard Business Review Press, Boston, Massachusetts (2005)
5. Liebowitz, J., Rubenstein-Montano, B., McCaw, D., et al.: The Knowledge Audit. Knowl Process Manag **7**, 3–10 (2000). doi:10.1002/(SICI)1099-1441(200001/03)7:1<3:AID-KPM72> 3.0.CO;2-0
6. Perez-Soltero, A., Barcelo-Valenzuela, M., Sanchez-Schmitz, G., Rodriguez-Elias, O.M.: A computer prototype to support knowledge audits in organizations. Knowl Process Manag **16**, 124–133 (2009). doi:10.1002/kpm.329
7. Velázquez Mendoza, M.J., Rodriguez-Elias, O.M., Rose Gómez, C.E., Meneses Mendoza, S. R.: Modelo para diseño de perfiles de conocimiento: una aplicacion en la industria generadora de energía eléctrica. Res Comput Sci **55**, 125–135 (2012)
8. Nonaka, I., Takeuchi, H.: The Knowledge-Creating Company: How Japanese Companies Create the Dynamics of Innovation. Oxford University Press (1995)

9. Teo, S.T.T., Lakhani, B., Brown, D., Malmi, T.: Strategic human resource management and knowledge workers: a case study of professional service firms. Manag Res News **31**, 683–696 (2008). doi:10.1108/01409170810898572

10. Ivancevich, J.M.: Administración de recursos humanos, 9a Edición. McGraw Hill, México, D.F (2004)

11. Bratton, J., Gold, J.: Human Resource Management: Theory and Practice, 2nd edn. Macmillan Press LTD (2000)

12. Drucker, P.F.: Management Challenges for the 21st Century. HarperBusiness (2001)

13. Wiig, K.M.: Knowledge Management: Where Did it Come from and Where Will it Go? Expert Syst. Appl. **13**, 13 (1997). doi:10.1016/S0957-4174(97)00018-3

14. Breaugh, J.A.: Research on employee recruitment: so many studies, so many remaining questions. J. Manag. **26**, 405–434 (2000). doi:10.1177/014920630002600303

15. Abraham, J., Morin, L., Renaud, S., et al.: What do experts expect from human resource practices. Glob. J. Bus. Res **7**, 121–133 (2013)

16. Lengnick-Hall, M.L., Lengnick-Hall, C.A., Andrade, L.S., Drake, B.: Strategic human resource management: the evolution of the field. Hum. Resour. Manag. Rev **19**, 64–85 (2009). doi:10.1016/j.hrmr.2009.01.002

17. Lewis, R.E., Heckman, R.J.: Talent management: a critical review. Hum. Resour. Manag. Rev **16**, 139–154 (2006). doi:10.1016/j.hrmr.2006.03.001

18. Saracho, J.M.: Talento organizacional. Un modelo para la definición organizacional del talento. RIL Editores, Santiago, Chile (2011)

19. Velázquez Mendoza, M.J., Rodriguez-Elias, O.M., Rose Gómez, C.E., Meneses Mendoza, S. R.: Perfiles de Conocimiento en la Gestión del Recurso Humano de las Organizaciones. Congr. Int. Invest. Acad. J. **4**, 3209–3214 (2012)

20. Martinsons, M.G.: Knowledge-based systems leverage human resource management expertise. Int. J. Manpower **16**, 17–35 (1995)

21. Guarino, N., Oberle, D., Staab, S.: What is an ontology? In: Staab, S., Studer, R. (eds.) Handbook Ontologies International Handbooks Information System, pp. 1–17. Springer (2009)

22. Gruber, T.R.: Toward principles for the design of ontologies used for knowledge sharing? Int. J. Hum. Comput. Stud. **43**, 907–928 (1995). doi:10.1006/ijhc.1995.1081

23. Brewster, C., O'Hara, K.: Knowledge representation with ontologies: present challenges-future possibilities. Int. J. Hum. Comput. Stud. **65**, 563–568 (2007). doi:10.1016/j.ijhcs.2007. 04.003

24. Houhamdi, Z., Athamena, B.: Ontology-based knowledge management. Int. J. Eng. Technol. **7**, 51–62 (2015). doi:10.1109/MC.2002.1046975

25. Rodríguez, O.M., Vizcaíno, A., Martínez, A.I., Piattini, M.: Using a multi-agent architecture to manage knowledge in the software maintenance process. In: 8th International Conference Knoweldge-Based Intelligent Information Engineering System (KES'04), LNCS(3213). Springer, pp 1181–1187 (2004)

26. Serna, M.E., Serna, A.A.: Ontology for knowledge management in software maintenance. Int. J. Inf. Manag. **34**, 704–710 (2014). doi:10.1016/j.ijinfomgt.2014.06.005

27. Vizcaíno, A., García, F., Piattini, M., Beecham, S.: A validated ontology for global software development. Comput. Stand. Interfaces **46**, 66–78 (2016). doi:10.1016/j.csi.2016.02.004

28. Fensel, D., Van Harmelen, F., Ding, Y., et al.: Ontology-based knowledge management. Computer (Long Beach Calif) **35**, 56–59 (2002). doi:10.1109/MC.2002.1046975

29. Rodríguez-Enríquez, C.A., Alor-Hernández, G., Mejia-Miranda, J., et al.: Supply chain knowledge management supported by a simple knowledge organization system. Electron. Commer. Res. Appl. **19**, 1–18 (2016). doi:10.1016/j.elerap.2016.06.004

30. Alor-Hernández, G., Sánchez-Ramírez, C., Cortes-Robles, G., et al.: BROSEMWEB: a brokerage service for e-procurement using semantic web technologies. Comput. Ind. **65**, 828–840 (2014). doi:10.1016/j.compind.2013.12.007

31. Armbrust, M., Stoica, I., Zaharia, M., et al.: A view of cloud computing. Commun. ACM **53**, 50 (2010). doi:10.1145/1721654.1721672

32. Rodríguez-García, M.Á., Valencia-García, R., García-Sánchez, F., Samper-Zapater, J.J.: Ontology-based annotation and retrieval of services in the cloud. Knowl.-Based Syst. **56**, 15–25 (2014). doi:10.1016/j.knosys.2013.10.006
33. Rodríguez-García, M.Á., Valencia-García, R., García-Sánchez, F., Samper-Zapater, J.J.: Creating a semantically-enhanced cloud services environment through ontology evolution. Future Gener. Comput. Syst. **32**, 295–306 (2014). doi:10.1016/j.future.2013.08.003
34. Paredes-Valverde, M.A., Rodríguez-García, M.Á., Ruiz-Martínez, A., et al.: ONLI: an ontology-based system for querying DBpedia using natural language paradigm. Expert Syst. Appl. **42**, 5163–5176 (2015). doi:10.1016/j.eswa.2015.02.034
35. Zhang, X., Hou, X., Chen, X., Zhuang, T.: Ontology-based semantic retrieval for engineering domain knowledge. Neurocomputing **116**, 382–391 (2013). doi:10.1016/j.neucom.2011.12.057
36. Fan, S., Hua, Z., Storey, V.C., Zhao, J.L.: A process ontology based approach to easing semantic ambiguity in business process modeling. Data Knowl. Eng. **102**, 57–77 (2016). doi:10.1016/j.datak.2016.01.001
37. Feilmayr, C., Wöß, W.: An analysis of ontologies and their success factors for application to business. Data Knowl. Eng. **101**, 1–23 (2016). doi:10.1016/j.datak.2015.11.003
38. Han, L., Chen, G., Li, M.: A method for the acquisition of ontology-based user profiles. Adv. Eng. Softw. **65**, 132–137 (2013). doi:10.1016/j.advengsoft.2013.06.008
39. Calegari, S., Pasi, G.: Personal ontologies: generation of user profiles based on the YAGO ontology. Inf. Process Manag. **49**, 640–658 (2013). doi:10.1016/j.ipm.2012.07.010
40. Luna, V., Quintero, R., Torres, M., et al.: An ontology-based approach for representing the interaction process between user profile and its context for collaborative learning environments. Comput. Hum. Behav. **51**, 1387–1394 (2015). doi:10.1016/j.chb.2014.10.004
41. Azevedo, C.L.B., Iacob, M.-E., Almeida, J.P.A., et al.: Modeling resources and capabilities in enterprise architecture: a well-founded ontology-based proposal for ArchiMate. Inf. Syst. **54**, 235–262 (2015). doi:10.1109/EDOC.2013.14
42. Avison, D.E., Lau, F., Myers, M.D., Nielsen, P.A.: Action research. Commun. ACM **42**, 94–97 (1999). doi:10.1145/291469.291479
43. Fernández-López, M., Gómez-Pérez, A., Juristo, N.: Methontology: from ontological art towards ontological engineering. In: Proceedings of the Ontological Engineering AAAI-97 Spring Symposium Series (1997)
44. Rodriguez-Elias, O.M., Rose-Gómez, C.E, Vizcaíno, A., Martínez-García, A.I.: Integrating current practices and information systems in KM initiatives: a knowledge management audit approach. In: International Conference Knowledge Management and Information Sharing 2nd International Joint Conference Knowledge Discovery Knowledge Engineering Knowledge Management INSTICC, Valencia, España, pp 71–80 (2010)
45. Farquhar, A., Fikes, R., Rice, J.: The ontolingua server: a tool for collaborative ontology construction. Int. J. Hum. Comput. Stud. **46**, 707–727 (1997). doi:10.1006/ijhc.1996.0121
46. Chaudhri, V., Farquhar, A., Fikes, R. et al.: Open knowledge base connectivity. Knowl. Syst. Lab. (1998)
47. Cañas, A.J., Novak, J.D.: Concept mapping using CmapTools to enhance meaningful learning. In: Sherborne, T., Okada, A., Buckingham Shum, S. (eds.) Knowledge Cartography Software Tools Mapping Technology, pp. 25–46. Springer, London (2008)
48. Protégé (2013) The protégé ontology editor and knowledge acquisition. http://protege.stanford.edu/
49. Chinosi, M., Trombetta, A.: BPMN: an introduction to the standard. Comput. Stand. Interfaces **34**, 124–134 (2012). doi:10.1016/j.csi.2011.06.002
50. Fernández-López, M., Gómez-Pérez, A., Juristo, N.: Methontology: from ontological art towards ontological engineering. In: Proceedings of the Ontological Engineering AAAI-97 Spring Symposium Series (1997)
51. The World Wide Web Consortium W3C OWG.: OWL 2 web ontology language document overview, 2nd edn (2012) Available at:http://www.w3.org/TR/owl2-overview

52. Horridge, M., Knublauch, H., Rector, A. et al.: A Practical Guide To Building OWL Ontologies Using Protégé 4 and CO-ODE Tools. The University Of Manchester (2009)
53. Shearer R, Motik B, Horrocks, I.: HermiT: a highly-efficient OWL reasoner. 5th International Workshop on OWL Experiences Directions (OWLED 2008 EU) (2008)
54. Rosas Daniel, J.A., Rodríguez-Elias, O.M., Velazquez-Mendoza, M.D.J., Rose-Gómez, C.E.: Diseño de un sistema para valoración de perfiles de recursos humanos. Rev. Coloq. Investig. Multidiscip. **3**, 403–414 (2015)

Chapter 4
Sentiment Analysis Based on Psychological and Linguistic Features for Spanish Language

María Pilar Salas-Zárate, Mario Andrés Paredes-Valverde, Miguel Ángel Rodríguez-García, Rafael Valencia-García and Giner Alor-Hernández

Abstract Recent research activities in the areas of opinion mining, sentiment analysis and emotion detection from natural language texts are gaining ground under the umbrella of affective computing. Nowadays, there is a huge amount of text data available in the Social Media (e.g. forums, blogs, and social networks) concerning to users' opinions about experiences buying products and hiring services. Sentiment analysis or opinion mining is the field of study that analyses people's opinions and mood from written text available on the Web. In this paper, we present extensive experiments to evaluate the effectiveness of the psychological and linguistic features for sentiment classification. To this purpose, we have used four psycholinguistic dimensions obtained from LIWC, and one stylometric dimension obtained from WordSmith, for the subsequent training of the SVM, Naïve Bayes, and J48 algorithms. Also, we create a corpus of tourist reviews from the travel website TripAdvisor. The findings reveal that the stylometric dimension is quite feasible for

M.P. Salas-Zárate (✉) · M.A. Paredes-Valverde · R. Valencia-García
Departamento de Informática y Sistemas, Universidad de Murcia,
Campus de Espinardo, 30100 Murcia, Spain
e-mail: mariapilar.salas@um.es

M.A. Paredes-Valverde
e-mail: marioandres.paredes@um.es

R. Valencia-García
e-mail: valencia@um.es

M.Á. Rodríguez-García
Computational Bioscience Research Center, King Abdullah University of Science
and Technology, 4700 KAUST, P.O. Box 2882, 23955-6900 Thuwal,
Kingdom of Saudi Arabia
e-mail: miguel.rodriguezgarcia@kaust.edu.sa

G. Alor-Hernández
Division of Research and Postgraduate Studies, Instituto Tecnológico de Orizaba,
Av. Oriente 9 no. 852 Col. E. Zapata, CP 94320 Orizaba Veracruz, Mexico
e-mail: galor@itorizaba.edu.mx

© Springer International Publishing AG 2017
G. Alor-Hernández and R. Valencia-García (eds.), *Current Trends
on Knowledge-Based Systems*, Intelligent Systems Reference Library 120,
DOI 10.1007/978-3-319-51905-0_4

sentiment classification. Finally, with regard to the classifiers, SVM provides better results than Naïve Bayes and J48 with an F-measure rate of 90.8%.

Keywords LIWC · Machine learning · Natural language processing · Opinion mining · Sentiment analysis

4.1 Introduction

With the booming of social media, some applications using sentiment analysis has been rapidly developed in recent years. The majority of people visit opinion sites, discussion forums or social networks aiming to obtain experiences of other users before taking a decision. Due to the number of reviews has exponentially increased on the Web, the task of reading all these opinions has become impossible for the users. For this reason, there is a need for automating the analysis and classification of these opinions.

Sentiment analysis (a.k.a. opinion mining) has become a popular topic in order to understand public opinion from unstructured Web data. From this perspective, sentiment analysis is devoted to extract users' opinions from textual data. The capture of public opinion is gaining momentum, particularly in terms of product preferences, marketing campaigns, political movements, financial aspects and company strategies. In this context, different techniques based on supervised learning and unsupervised learning, may be used. Thanks to these techniques, several attempts at sentiment classification were done. For instance, the problems that have been studied are mainly focused around subjectivity identification [1], analysis and building sentiment lexicons [2, 3], evaluation and classification of the Twitter messages [4–6], features selection [7], negation [8]. Some other proposals have tried to introduce sentiment classification problem in different levels i.e. document-level [9, 10], sentence-level [11], word-level and aspect-level [12]. However, one of the main issues on sentiment classification is the existence of many conceptual rules that govern the linguistic expression of sentiments. Human psychology, which relates to social, cultural, personal and biological aspects, can be an important feature to be considered in sentiment analysis. For this purpose, the LIWC and WordSmith text analysis software are useful tools since they enable the extraction of psychological and linguistic features from natural language text.

On the other hand, it is worth mentioning that most sentiment analysis studies deal exclusively with English documents, perhaps owing to the lack of resources in other languages. Considering that the Spanish language has a much more complex syntax than many other languages, and that it is the third most widely spoken language in the world, we firmly believe that the computerization of Internet domains in this language is of utmost importance.

The main contribution of this work is the evaluation of the effectiveness of different psychological and linguistic features for sentiment classification in the Spanish language. This process covered five categories (positive, negative, neutral, highly positive and highly negative) and it was performed by using different classifiers. Also, a corpus of tourism reviews, obtained from the travel website TripAdvisor, was built and compiled.

This paper is structured as follows. Section 4.2 presents the state of the art on opinion mining and sentiment analysis. Section 4.3 presents the tourism corpus used along the evaluation performed. Section 4.4 describes the LIWC's dimensions and stylometric variables provided by Wordsmith. Section 4.5 presents the classification algorithms SVM (Support Vector Machine), Naïve Bayes, and J48 used in this work. Section 4.6 provides a detailed description about the experiments performed. Section 4.7 presents the results of the classifiers' evaluation performed by using LIWC's dimensions and stylometric variables described in Sects. 4.4 and 4.5. Section 4.8 discusses the results obtained by our approach, and a comparison is presented with two machine-learning based natural language processing (NLP) tools, the OpenNLP and Stanford Classifier. Finally, Sect. 4.9 describes conclusions and future work.

4.2 Related Work

In recent years, several proposals have presented studies based on two main approaches in the polarity classification: Semantic Orientation and Machine learning.

The Semantic Orientation (SO) approach uses lexical resources such as lexicons, which have been automatically or semi-automatically generated [13–15]. These lexicons have usually extended the WordNet lexical database. Two clear examples of this kind of lexicons are WordNet-Affect [14] and SentiWordNet [13].

In the literature, several proposals based on sentiment lexicons have been presented. For instance, in [16] a lexicon-based approach to extract sentiment from texts was proposed. This approach is based on The Semantic Orientation CALculator (SO-CAL), which uses dictionaries of words annotated with their semantic orientation (polarity and strength), and incorporates intensification and negation.

Peñalver-Martinez et al. [12] proposed an innovative opinion mining methodology that takes advantage of new semantic Web-guided solutions to enhance the results obtained with traditional natural language processing techniques and sentiment analysis processes and Semantic Web technologies. Concretely, the proposed approach is based on three different stages: (1) an ontology based mechanism for feature identification, (2) a technique to assign a polarity to each feature based on SentiWordNet and (3) a new approach for opinion mining based on vector analysis. Ghosh and Animesh [17] presented a rule-based method to identify the sentiment polarity of opinion sentences. SentiWordNet was used to calculate overall sentiment score of each sentence. Also, the results showed that SentiWordNet could be used as an important resource for sentiment classification tasks.

Most studies on opinion mining have been concerned to the English language. Indeed, a main disadvantage of the approach based on sentiment lexicons is the lack of sentiment lexicons for non-English languages, therefore it is difficult its application to other languages. Despite the aforementioned issue, there are some interesting works that provide sentiment lexicons oriented to languages such as Spanish [15, 18], German [19], Dutch [20] and Arabic [21].

Also, due to the limitation mentioned above, some proposals have used alternative methods. The first alternative method uses some lexical resources for sentiment analysis based on WordNet (regardless of the target language, for example, the English language) jointly with a lexical database based on WordNet and specific for the target language or multilingual [22].

The second alternative method uses a different sentiment lexicon from the target language. Therefore, an automatic translation of the corpus is carried out before of the polarity classification phase. However, this method depends on the availability and reliability of the automatic translation engine available [23, 24].

On the other hand, there are some proposals [25–28] based on psycholinguistic tools for sentiment analysis such as LIWC, a tool that offers a dictionary in several languages such as: Spanish, English, French, and German, among others. These proposals construct a semantic orientation-based lexicon base on two ("Positive emotion" and "Negative emotion") of the 76 categories of LIWC. These categories include words such as love, nice, good, great, hurt, ugly, sad, bad, and worse, among others. The LIWC's dictionaries are used on different studies covering languages such as English [25–28], Portuguese [26] and Spanish [29].

Furthermore, with regard to the Machine learning approach, it is worth mentioning that this approach often relies on supervised classification approaches. These approaches use a collection of data to train the classifier algorithms. Among the machine learning techniques commonly used in the sentiment polarity classification are Support Vector Machine (SVN), Naive Bayes (NB), Maximum Entropy (MaxEnt), among others. For example, in [30] compared three supervised machine learning algorithms (Naïve Bayes, SVM and the character based N-gram model) for sentiment classification. This comparison was performed on user's reviews obtained from travel blogs. Authors concluded that well-trained machine learning algorithms can provide a very good performance for sentiment polarity classification on the travel domain. Sidorov et al. [31] explored how different settings (n-gram size, corpus size, number of sentiment classes, balanced vs. unbalanced corpus, various domains) affect the precision of the machine learning algorithms. Naïve Bayes, Decision Tree, and Support Vector Machines were considered.

It is necessary to remark that this work differs from the works presented above in several aspects: (i) our study consider all LIWC's categories for the sentiment analysis. Although the LIWC dictionary has been evaluated in different works for the sentiment analysis, only the "Positive emotion" and "negative emotion" categories have been considered for constructing a lexicon based on semantic orientation, (ii) we have evaluated the combination of psychological and linguistic features extracted by using the LIWC and WordSmith text analyzers, (iii) we carried out a comparison of our approach with two sentiment analysis tools (Stanford Classifier and OpenNLP) based on machine learning, and (iv) unlike of several works, which use only two classes (positive and negative), our approach uses five classes (positive, negative, neutral, highly positive and highly negative).

This was done in order to find out how the number of classes affects the precision of the classification algorithms.

4.3 Corpus

For this study, we have chosen the TripAdvisor® website since is the world's largest travel site with 315 million unique users per month and over 200 million reviews and opinions of over 4.4 million accommodations, restaurants and attractions.

The corpus consists of 1600 reviews, obtained from the TripAdvisor® website, concerning hotels, restaurants, and museums, among other topics. Each review was examined and classified by hand in order to ensure the quality of the corpus. Also, a value, from 1 to 5, was assigned to each review according to next criteria:

- 5, when the review has a highly positive sentiment.
- 4, when the review has a positive sentiment.
- 3, when the review has no sentiment.
- 2, when the review has a negative sentiment.
- 1, when the review has a highly negative sentiment.

The classification results were stored in an XML-based format. Figure 4.1 shows an example of a review with highly positive orientation.

Table 4.1 shows the total amount of reviews classified in each category mentioned above. It should be mentioned that the aforementioned activities were performed along six months by a group of three experts in order to ensure the quality of the corpus.

```
<review id="3" rank="5" topic="Alojamiento">
    <abstract>
        Gran hallazgo
    </abstract>
    <content>
        Fantástico hotel, personal muy amable, habitaciones espaciosas, desayuno excelente
        y muy silencioso para descansar. El vecindario es uno de los mejores de Barna, el
        jardín en la zona posterior es una delicia. Muy pero muy recomendable.
    </content>
</review>
```

Fig. 4.1 Example of a review with highly positive orientation

Table 4.1 Classification of the corpus

Value	Classification	Number of reviews
5	Highly positive	320
4	Positive	320
3	Neutral	320
2	Negative	320
1	Highly negative	320
	Total	1600

Table 4.2 Reviews classified into three categories

Value	Classification	Number of reviews
5	Positive	640
4		
3	Neutral	320
2	Negative	640
1		
	Total	1600

Table 4.3 Reviews classified into two categories

Value	Classification	Number of reviews
5	Positive	640
4		
2	Negative	640
1		
	Total	1280

In order to find out how the number of classes affects the precision of the classification algorithms, the corpus was also classified into three categories (Positive, Negative and Neutral). Based on this understanding, the reviews with a 4 and 5 value were classified as positive. Meanwhile, the reviews with a 1 and 2 value were classified as negative. The results of this classification are shown in Table 4.2.

Finally, in order to evaluate the classifier algorithms with only two categories, the neutral opinions obtained from the aforementioned process were omitted. The results of this process are shown in Table 4.3.

4.4   LIWC and Stylometric Variables

LIWC (Linguistic Inquiry and Word Count) is a software application that provides an effective tool for studying the emotional, cognitive, and structural components contained in language on a word-by-word basis. Early approaches to psycholinguistic concerns involved almost exclusively qualitative philosophical analyses. In this field, more modern researches have provided empirical evidence on the relation between language and the state of mind of subjects, or even their mental health. In this regard, further studies such as [32] have dealt with the therapeutic effect of verbally expressing emotional experiences and memories. LIWC was developed for providing an efficient method for studying these psycholinguistic concerns thanks to corpus analysis, and it has been considerably improved since its first version [33]. An updated revision of the original application was presented in [34], namely LIWC2001.

For this study, we used the LIWC2007 Spanish dictionary, which is composed by 7515 words and word stems. Each word is classified into one or more of the 72 categories included by default in LIWC. Also, these categories are classified into four dimensions: (1) standard linguistic process, (2) psychological process, (3) relativity, and (4) personal concerns.

Table 4.4 shows some examples of the LIWC categories. The full list of categories is presented in [35].

On the other hand, we have used further stylometric variables, due to that there are some linguistic features not included in LIWC standard linguistic dimensions, which has been considered significant for this study. We used word level measures such as: word length distribution, average number words per sentence, word length distribution containing frequency of 1 letter word to frequency of 14 letters word. These linguistic features were obtained from the WordSmith software.

4.5 Machine Learning Approach

In a classification-based machine learning approach, two sets of data are required: training and validation set. The first set is used by an automatic classifier to learn the differentiating characteristics of documents, and the second set is used to validate the performance of the automatic classifier. Machine learning classifiers such as Naive Bayes (NB), Maximum Entropy (ME), and Support Vector Machines (SVM) have been used on different works achieving great success in text categorization [36–38].

For this work, we used WEKA [39] to evaluate the classification success of reviews (positive, negative, neutral, highly positive or highly negative) based on LIWC and WordSmith categories.

WEKA provides several classification algorithms (also called classifiers), which allows the creation of models according to the data and purpose of analysis. Classification algorithms are categorized into seven groups: Bayesian (Naïve Bayes, Bayesian nets, etc.), functions (linear regression, SMO, logistic, etc.), lazy (IBk, LWL, etc.), meta-classifiers (Bagging, Vote, etc.), miscellaneous (Serialized Classifier and InputMappedClassifier), rules (DecisionTable, OneR, etc.) and trees (J48, RandomTree, etc.). The classification process involves the building of a model based on the analysis of the instances. This model is represented through classification rules, decision trees, or mathematical formulae. The model is used to generate the classification of unknown data, calculating the percentage of instances which were correctly classified.

Table 4.4 LIWC dimensions

Dimension	Categories
1. Linguistic processes	Word count, negations, articles, prepositions, among others
2. Psycological processes	Positive emotions, negative emotions, feeling, communication, among others
3. Relativity	Past tense verb, future tense verb, inclusive, among others
4. Personal concerns	Job or work, sports, music, religion, among others

4.6 Experiment

This section presents the experiment conducted to study the effectiveness of the psychological and linguistic features for sentiment classification into five categories: highly positive, highly negative, positive, negative and neutral. As it will be shown in next sections, the approach has been tested on the Spanish language by using a corpus of reviews concerning to the tourism domain. Figure 4.2 shows the general process of the experiment.

4.6.1 Combination of LIWC Dimensions and Stylometric Dimension

Aiming to detect the set of dimensions that provide the best results and to validate if the stylometric dimension improves the success rates, we have obtained 11 combinations of LIWC dimensions, and 15 for all possible combinations of stylometric dimension and LIWC dimensions (see Table 4.5).

4.6.2 Text Analysis with LIWC and WordSmith

The tourism corpus was analyzed with LIWC and WordSmith software through all the possible combinations presented in Table 4.5 and taking into account three possible sets of opinion classes (positive-negative, positive-neutral-negative and highly positive-positive-neutral-negative-highly negative).

LIWC searches for target words or word stems from the dictionary, categorizes them into one of its linguistic dimensions, and then converts the raw counts to

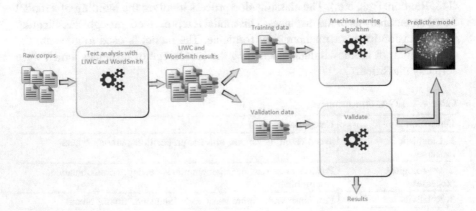

Fig. 4.2 General process of the experiment

Table 4.5 Combination of LIWC dimensions and stylometric dimension

Dimensions	Description
1	Dimension 1 LIWC
2	Dimension 2 LIWC
3	Dimension 3 LIWC
4	Dimension 4 LIWC
1-2	Combination of LIWC dimensions 1 and 2
1-3	Combination of LIWC dimensions 1 and 3
1-4	Combination of LIWC dimensions 1 and 4
2-3	Combination of LIWC dimensions 2 and 3
2-4	Combination of LIWC dimensions 2 and 4
3-4	Combination of LIWC dimensions 1 and 2
1-2-3	Combination of LIWC dimensions 1, 2 and 3
1-2-4	Combination of LIWC dimensions 1, 2 and 4
1-3-4	Combination of LIWC dimensions 1, 3 and 4
2-3-4	Combination of LIWC dimensions 2, 3 and 4
1-2-3-4	Combination of all LIWC dimensions
Styl.	Further stylometric variables
1 + Styl.	Combination LIWC dimension 1 and Further stylometric variables
2 + Styl.	Combination LIWC dimension 2 and Further stylometric variables
3 + Styl.	Combination LIWC dimension 3 and Further stylometric variables
4 + Styl.	Combination LIWC dimension 4 and Further stylometric variables
1_2 + Styl.	Combination LIWC dimension 1, 2 and Further stylometric variables
1_3 + Styl.	Combination LIWC dimension 1, 3 and Further stylometric variables
1_4 + Styl.	Combination LIWC dimension 1, 4 and Further stylometric variables
2_3 + Styl.	Combination LIWC dimension 2, 3 and Further stylometric variables
2_4 + Styl.	Combination LIWC dimension 2, 4 and Further stylometric variables
3_4 + Styl.	Combination LIWC dimension 3, 4 and Further stylometric variables
1_2_3 + Styl.	Combination LIWC dimension 1, 2, 3 and Further stylometric variables
1_2_4 + Styl.	Combination LIWC dimension 1, 2, 4 and Further stylometric variables
1_3_4 + Styl.	Combination LIWC dimension 1, 3, 4 and Further stylometric variables
2_3_4 + Styl.	Combination LIWC dimension 2, 3, 4 and Further stylometric variables
1_2_3_4 + Styl.	Combination all LIWC dimension and Further stylometric variables

percentages of total words. On the other hand, WordSmith generates statistical information such as average sentence and average word length, among others.

4.6.3 Training a Machine Learning Algorithm and Validation Test

A set of the results obtained by the analysis with LIWC and WordSmith was used in order to train the machine learning algorithms, which are subsequently validated with the remaining subset. For the experiment, J48, Naive Bayes (NB), and the SMO algorithm for SVM classifiers were used.

Specifically, for J48, NB, and SMO classifiers, a ten-fold cross-validation was done. This technique is used to evaluate how the results obtained would generalize to an independent data set. Since the aim of this experiment is the prediction of the positive, negative, neutral, highly positive and highly negative condition of the texts, a cross-validation was applied in order to estimate the accuracy of the predictive models. It involved partitioning a sample of data into complementary subsets, performing an analysis on the training set and validating the analysis on the testing or validation set.

4.7 Evaluation and Results

In order to measure the performance of our method, we have used three evaluation measurements that are commonly used in sentiment analysis: precision, recall and F-measure. Recall (1) is the proportion of actual positive cases that were correctly predicted as such. On the other hand, precision (2) represents the proportion of predicted positive cases that are real positives. Finally, F-measure (3) is the harmonic mean of precision and recall.

(1) Recall = TP/(TP + FN)
(2) Precision = TP/(TP + FP)
(3) F1 = 2 *(Precision * Recall)/(Precision + Recall)

4.7.1 Results for the Tourism Corpus

This section describes the results obtained of each possible combination described in the previous section. Tables 4.6, 4.7 and 4.8 present the evaluation results by using two (positive-negative), three (positive-neutral-negative) and five (highly positive-positive-neutral-negative-highly negative) categories respectively. In the first column, the number of LIWC dimensions used for each classifier is indicated. For instance, 1_2_3_4 indicates that all the dimensions have been used in the experiment, and 1_2 indicates that only the 1 and 2 dimensions have been used to train the classifier.

Table 4.6 shows that the SMO algorithm provides better results than J48 and Naïve Bayes. When the evaluation involved only one dimension, the J48 and Naïve algorithms obtained the best results through the first dimension (Linguistic processes). On the other hand, the SMO algorithm obtained the best result through the second dimension (Psychological processes). Conversely, the fourth dimension (Personal concerns) provided the worst results for the three classification algorithms. On the other hand, the combination of all LIWC dimensions and the stylometric

Table 4.6 Classification results obtained by using two classes (positive-negative)

	J48			Naïve Bayes			SMO		
	P	R	F1	P	R	F1	P	R	F1
1	0.83	0.83	0.83	0.809	0.802	0.799	0.825	0.823	0.823
2	0.788	0.788	0.787	0.808	0.791	0.784	0.836	0.834	0.834
3	0.737	0.735	0.734	0.728	0.702	0.68	0.736	0.736	0.736
4	0.667	0.661	0.651	0.655	0.654	0.652	0.673	0.66	0.633
1_2	0.806	0.805	0.805	0.838	0.833	0.831	0.898	0.898	0.898
1_3	0.799	0.798	0.798	0.804	0.804	0.804	0.833	0.832	0.832
1_4	0.8	0.799	0.799	0.785	0.78	0.778	0.836	0.834	0.833
2_3	0.803	0.803	0.803	0.8	0.787	0.781	0.852	0.851	0.85
2_4	0.8	0.798	0.798	0.787	0.782	0.78	0.837	0.836	0.836
3_4	0.744	0.744	0.743	0.693	0.69	0.687	0.735	0.734	0.734
1_2_3	0.821	0.821	0.821	0.829	0.822	0.82	0.905	0.905	0.905
1_2_4	0.819	0.819	0.819	0.831	0.829	0.828	0.905	0.905	0.905
1_3_4	0.785	0.784	0.784	0.772	0.772	0.772	0.829	0.828	0.828
2_3_4	0.767	0.767	0.767	0.79	0.784	0.782	0.851	0.85	0.85
1_2_3_4	0.818	0.818	0.818	0.825	0.822	0.821	0.903	0.903	0.903
Styl.	0.697	0.697	0.697	0.759	0.719	0.693	0.749	0.743	0.739
1 + Styl.	0.77	0.77	0.77	0.788	0.756	0.741	0.828	0.824	0.823
2 + Styl.	0.772	0.772	0.772	0.788	0.775	0.799	0.86	0.859	0.859
3 + Styl.	0.718	0.717	0.717	0.758	0.725	0.704	0.772	0.772	0.772
4 + Styl.	0.688	0.688	0.688	0.753	0.72	0.697	0.752	0.751	0.75
1_2 + Styl.	0.81	0.81	0.81	0.829	0.819	0.816	0.902	0.902	0.902
1_3 + Styl.	0.766	0.766	0.765	0.796	0.768	0.756	0.83	0.828	0.828
1_4 + Styl.	0.772	0.772	0.772	0.783	0.756	0.743	0.833	0.83	0.829
2_3 + Styl.	0.773	0.773	0.773	0.794	0.785	0.782	0.869	0.869	0.869
2_4 + Styl.	0.785	0.785	0.785	0.785	0.773	0.767	0.859	0.859	0.859
3_4 + Styl.	0.706	0.706	0.706	0.753	0.726	0.708	0.774	0.773	0.773
1_2_3 + Styl.	0.816	0.816	0.816	0.83	0.822	0.819	0.906	0.906	0.906
1_2_4 + Styl.	0.798	0.798	0.798	0.82	0.809	0.806	0.907	0.907	0.907
1_3_4 + Styl.	0.765	0.765	0.765	0.787	0.764	0.754	0.829	0.828	0.828
2_3_4 + Styl.	0.788	0.788	0.788	0.791	0.783	0.779	0.865	0.864	0.864
1_2_3_4 + Styl.	0.806	0.806	0.806	0.822	0.813	0.811	0.908	0.908	0.908

dimension provided the best classification result with an F measure of 90.8%. As can be seen, the addition of the stylometric dimension improved the classification results.

The classification results obtained by using three classes are presented in Table 4.7. We found that the SMO classification algorithm provides the best results. Individually, the first dimension (Linguistic processes) provided the best results for the J48 and Naïve Bayes algorithms. On the other hand, the second dimension (Psychological processes) provided the best results for the SMO

Table 4.7 Classification results obtained by using three classes (positive-neutral-negative)

	J48			Naïve Bayes			SMO		
	P	R	F1	P	R	F1	P	R	F1
1	0.724	0.744	0.731	0.712	0.741	0.713	0.66	0.774	0.71
2	0.679	0.693	0.685	0.708	0.732	0.7	0.675	0.793	0.727
3	0.655	0.689	0.653	0.631	0.66	0.611	0.604	0.704	0.648
4	0.619	0.638	0.598	0.615	0.614	0.614	0.581	0.653	0.583
1_2	0.715	0.721	0.718	0.746	0.763	0.748	0.792	0.842	0.776
1_3	0.698	0.702	0.7	0.713	0.724	0.718	0.667	0.783	0.718
1_4	0.706	0.718	0.708	0.714	0.699	0.702	0.668	0.784	0.718
2_3	0.697	0.702	0.699	0.704	0.714	0.691	0.684	0.804	0.737
2_4	0.684	0.689	0.686	0.716	0.711	0.707	0.672	0.789	0.724
3_4	0.636	0.646	0.64	0.631	0.616	0.619	0.614	0.716	0.658
1_2_3	0.724	0.721	0.722	0.742	0.746	0.739	0.798	0.842	0.78
1_2_4	0.716	0.72	0.718	0.756	0.748	0.749	0.802	0.845	0.784
1_3_4	0.696	0.701	0.698	0.714	0.691	0.7	0.672	0.789	0.723
2_3_4	0.682	0.688	0.684	0.708	0.694	0.694	0.706	0.804	0.74
1_2_3_4	0.725	0.73	0.73	0.758	0.744	0.747	0.775	0.839	0.783
Styl.	0.625	0.632	0.628	0.663	0.681	0.625	0.616	0.714	0.654
1 + Styl.	0.674	0.678	0.676	0.687	0.691	0.657	0.662	0.776	0.711
2 + Styl.	0.679	0.682	0.68	0.696	0.713	0.69	0.688	0.809	0.742
3 + Styl.	0.621	0.628	0.624	0.666	0.678	0.633	0.632	0.741	0.681
4 + Styl.	0.616	0.623	0.619	0.678	0.673	0.644	0.621	0.724	0.665
1_2 + Styl.	0.721	0.716	0.718	0.731	0.738	0.724	0.822	0.843	0.779
1_3 + Styl.	0.653	0.656	0.654	0.703	0.698	0.675	0.669	0.786	0.721
1_4 + Styl.	0.678	0.677	0.677	0.693	0.681	0.662	0.673	0.789	0.724
2_3 + Styl.	0.678	0.674	0.676	0.706	0.719	0.704	0.691	0.813	0.745
2_4 + Styl.	0.68	0.679	0.679	0.71	0.703	0.696	0.686	0.807	0.74
3_4 + Styl.	0.622	0.622	0.622	0.679	0.671	0.653	0.631	0.739	0.679
1_2_3 + Styl.	0.726	0.723	0.725	0.733	0.734	0.726	0.791	0.837	0.779
1_2_4 + Styl.	0.717	0.718	0.717	0.732	0.721	0.718	0.789	0.841	0.783
1_3_4 + Styl.	0.666	0.664	0.665	0.704	0.686	0.675	0.674	0.79	0.725
2_3_4 + Styl.	0.663	0.662	0.663	0.714	0.706	0.704	0.731	0.812	0.749
1_2_3_4 + Styl.	0.724	0.726	0.725	0.743	0.729	0.73	0.803	0.844	0.792

algorithm. Quite the reverse, the third and four dimensions provide the worst results. Furthermore, the combination of all LIWC dimensions and the stylometric dimension provided the best classification result with an F-measure of 79.2%. As can be seen, in these results also the addition of the stylometric dimension improved the classification results.

Table 4.8 Classification results obtained by using five classes (highly positive-positive-neutral-negative-highly negative)

	J48			Naïve Bayes			SMO		
	P	R	F1	P	R	F1	P	R	F1
1	0.558	0.558	0.558	0.579	0.586	0.572	0.611	0.61	0.603
2	0.566	0.566	0.566	0.58	0.588	0.562	0.607	0.615	0.605
3	0.534	0.534	0.534	0.529	0.539	0.507	0.537	0.557	0.522
4	0.54	0.539	0.538	0.52	0.518	0.498	0.538	0.533	0.512
1_2	0.598	0.599	0.598	0.59	0.603	0.582	0.653	0.66	0.65
1_3	0.578	0.579	0.578	0.572	0.588	0.568	0.609	0.611	0.607
1_4	0.561	0.561	0.561	0.566	0.559	0.549	0.61	0.606	0.603
2_3	0.589	0.591	0.59	0.58	0.586	0.564	0.624	0.633	0.622
2_4	0.585	0.586	0.586	0.577	0.574	0.568	0.61	0.616	0.609
3_4	0.536	0.537	0.536	0.529	0.522	0.515	0.543	0.554	0.542
1_2_3	0.577	0.575	0.575	0.586	0.599	0.579	0.636	0.644	0.637
1_2_4	0.575	0.576	0.575	0.605	0.603	0.601	0.647	0.652	0.646
1_3_4	0.566	0.566	0.566	0.566	0.563	0.559	0.6	0.601	0.598
2_3_4	0.575	0.576	0.575	0.583	0.578	0.574	0.621	0.629	0.62
1_2_3_4	0.601	0.603	0.602	0.594	0.596	0.592	0.645	0.651	0.645
Styl.	0.539	0.539	0.539	0.509	0.543	0.479	0.551	0.564	0.549
1 + Styl.	0.59	0.59	0.59	0.543	0.561	0.521	0.61	0.609	0.601
2 + Styl.	0.575	0.576	0.576	0.556	0.584	0.55	0.606	0.619	0.608
3 + Styl.	0.543	0.544	0.544	0.501	0.541	0.496	0.556	0.575	0.555
4 + Styl.	0.543	0.543	0.543	0.553	0.543	0.524	0.552	0.561	0.553
1_2 + Styl.	0.601	0.602	0.601	0.576	0.594	0.566	0.652	0.661	0.651
1_3 + Styl.	0.583	0.583	0.583	0.543	0.564	0.529	0.597	0.603	0.596
1_4 + Styl.	0.566	0.566	0.566	0.557	0.553	0.542	0.608	0.608	0.604
2_3 + Styl.	0.575	0.576	0.575	0.549	0.576	0.547	0.615	0.626	0.616
2_4 + Styl.	0.57	0.571	0.57	0.577	0.581	0.572	0.61	0.617	0.611
3_4 + Styl.	0.538	0.538	0.537	0.546	0.544	0.53	0.552	0.569	0.554
1_2_3 + Styl.	0.599	0.599	0.599	0.565	0.587	0.561	0.64	0.648	0.639
1_2_4 + Styl.	0.578	0.578	0.578	0.578	0.586	0.576	0.645	0.65	0.644
1_3_4 + Styl.	0.595	0.595	0.595	0.556	0.558	0.547	0.6	0.603	0.597
2_3_4 + Styl.	0.554	0.555	0.554	0.58	0.588	0.579	0.594	0.605	0.596
1_2_3_4 + Styl.	0.592	0.593	0.592	0.589	0.589	0.589	0.624	0.631	0.624

Table 4.8 shows that the SMO algorithm obtained better results than J48 and Naïve Bayes when five classes were involved. Individually, the second dimension (Psychological processes) provides the best result through the J48 and SMO algorithms. On the other hand, the Naïve Bayes algorithm obtained the best result through the first dimension (Linguistic processes). Conversely, the third and four

dimensions provided the worst results. On the other hand, the combination of the first and second LIWC dimensions and the stylometric dimension provided the best classification results with an F-measure of 65.1%. In these results, the third and fourth dimensions are least revealing.

4.8   Discussion of Results

The classification results obtained show that reducing the number of classes the classifiers precision increases i.e. the classification with two categories (positive-negative) (see Fig. 4.3) provided better results than the classification with three (positive-neutral-negative) (see Fig. 4.4) and five (highly positive-positive-neutral-negative-highly negative) (see Fig. 4.5) categories. Thus, it is by virtue of the combination of fewer categories that the classification algorithm performs a better classification, probably due to the fact that in a bipolar system there is less space for the classification of slippery cases. It also means that additional criteria and features are required to get a fine-grained classification into 5 categories for instance.

With regard to the classification algorithms, SMO Algorithm for SVM classifier obtained the best results. This algorithm has proved to be simpler, easier to implement, generally faster. Also, these results can be justified by the analysis presented in [40], where it is clearly shown how SVM models are more accurate in comparison to other classification algorithms such as: decision trees, neural

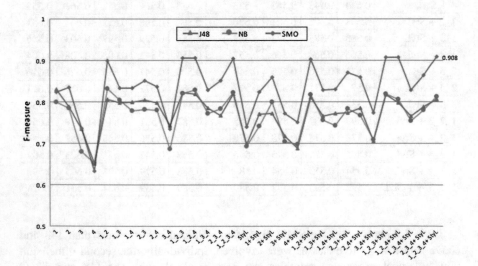

Fig. 4.3 Evaluation results for SMO, NB, J48, by using two classes

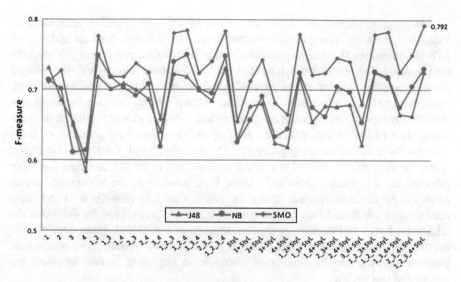

Fig. 4.4 Evaluation results for SMO, NB, J48, by using three classes

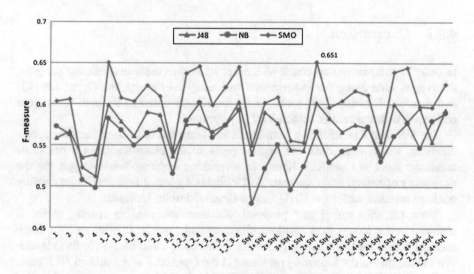

Fig. 4.5 Evaluation results for SMO, NB, J48, by using five classes

network, Bayesian network, nearest neighbor. SVM algorithm has been success-fully applied for text classification due to its main advantages: firstly, it has the ability to generalize well in high dimensional feature spaces; secondly, SVM eliminates the need for feature selection, making the application of text catego-rization considerably easier. Finally, another advantage of SVM algorithm over the conventional methods is its robustness [41].

Also, results showed that the combination of different LIWC dimensions provided better results than individual dimensions. Individually, the first and second LIWC dimensions (Linguistic processes and psychological processes) provided the best results. The first dimension contains grammatical words and the second dimension words related to the psychological process, such as: positive feelings, optimism, anxiety, sadness or depression, among others. Written opinions frequently contain a great amount of grammatical words and words related to emotional state of the author, which confirming the discriminatory potential of these dimensions in classification experiments. On the other hand, the fourth dimension (personal concerns) provided the worst results, owing to the fact that the topic selected for this study, "Tourism", bears little relation to the vocabulary corresponding to these categories. It can be stated that this dimension is the most content-dependent, and thus the least revealing. Also, general results show that the addition of the stylometric dimension improved the classification results. We ascribe this to the fact that most of the negative reviews contain more words than positive reviews, i.e. the number of words is an important feature to detect the polarity of the review.

4.8.1   Comparison

In order to compare our approach with other NLP tools used on sentiment analysis, the corpus used along our experiments was analyzed through the OpenNLP [42] and Stanford Classifier [43] tools. Then, the results obtained by both tools were compared with the results obtained by our approach.

The Stanford Classifier is a Java-based implementation of a maximum entropy classifier, which takes data and applies probabilistic classification. On the other hand, the Apache OpenNLP library is a machine learning based toolkit for the processing of natural language text. This includes a general Java maximum entropy package released under the GNU Lesser General Public License.

Table 4.9 shows that our proposal obtained encouraging results. Also, it obtained better results than Stanford classifier and OpenNLP, with an F-measure score of 90.8% for two classes, 79.2% for three classes, and 65.1% for five classes. We ascribe this to the following reasons: (1) the OpenNLP and Stanford NLP tools are based on maximum entropy models. Despite that Maximum entropy classifier provides good results in several proposals, SVM classifier outperforms the MaxEnt classifier in terms of the accuracy [44–46]. It can be justified for the SVM models advantages mentioned above in section "Discussion of results", and (2) unlike of these tools, our approach is based on the psycholinguistic features extraction to train the classifier. Therefore, we consider that these features are determinant in the classification results.

Table 4.9 Comparison of Standford and OpenNLP with our proposal

	Pos-Neg			Pos-Neu-Neg			highlyPos-Pos-Neu-Neg-highlyNeg		
	P	R	F1	P	R	F1	P	R	F1
Our approach	0.908	0.908	0.908	0.803	0.844	0.792	0.652	0.661	0.651
Stanford classifier	0.847	0.847	0.847	0.77	0.77	0.77	0.611	0.602	0.602
OpenNLP	0.853	0.849	0.848	0.782	0.75	0.728	0.647	0.65	0.646

4.9 Conclusion and Future Work

In this piece of research, we presented the feasibility study of LIWC dimensions and stylometric dimension in the sentiment classification.

In order to conduct a comprehensive study, we considered two, three and five categories (positive-negative, positive-neutral-negative and highly positive-positive-neutral-negative-highly negative) for the classification of touristic reviews in Spanish language. Also, three classifiers (SVM, NB, and J48) were used in order to evaluate the efficacy of psychological and linguistic features for sentiment classification. The results showed that the classification of reviews with two categories "positive-negative" provides better results than with other categories. Also, we found that the best results were obtained with the SMO classifier.

Additionally, the findings reveal that the stylometric dimension improved the classification results. This fact indicates that these features are quite feasible for sentiment classification. Finally, we compared our approach with two natural NLP tools (OpenNLP and standford classifier) based on machine-learning. The performance was measured by using the precision, recall, and F-measure metrics. Our proposal obtained encouraging results with a high F-measure score of 90.8%.

With regard to future research, we consider to evaluate our approach by using new corpora concerning to different domains such as financial and movies domain. Also, we plan to verify the efficiency of our approach, i.e., the use of LIWC and stylometric dimensions on sentiment analysis, on non-Spanish languages such as French, English, Italian and German. Finally, we also attempt to apply this approach for detecting the satire in Twitter messages.

Acknowledgements This work has been partially supported by the Spanish Ministry of Economy and Competitiveness and the European Commission (FEDER/ERDF) through project KBS4FIA (TIN2016-76323-R). María Pilar Salas-Zárate and Mario Andrés Paredes-Valverde are supported by the National Council of Science and Technology (CONACYT), and the Mexican government.

References

1. Abdul-Mageed, M., Diab, M., Kübler, S.: SAMAR: Subjectivity and sentiment analysis for Arabic social media. Comput. Speech Lang. **28**(1), 20–37 (2014)
2. Huang, S., Niu, Z., Shi, C.: Automatic construction of domain-specific sentiment lexicon based on constrained label propagation. Knowl.-Based Syst. **56**, 191–200 (2014)
3. Hogenboom, A., Heerschop, B., Frasincar, F., Kaymak, U., de Jong, F.: Multi-lingual support for lexicon-based sentiment analysis guided by semantics. Decis. Support Syst. **62**, 43–53 (2014)
4. Bae, Y., Lee, H.: Sentiment analysis of twitter audiences: measuring the positive or negative influence of popular twitterers. J. Am. Soc. Inf. Sci. Technol. **63**(12), 2521–2535 (2012)
5. Montejo-Ráez, A., Martínez-Cámara, E., Martín-Valdivia, M.T., Ureña-López, L.A.: A knowledge-based approach for polarity classification in Twitter. J. Assoc. Inf. Sci. Technol. **65**(2), 414–425 (2014)
6. Singhal, K., Agrawal, B., Mittal, N.: Modeling Indian general elections: sentiment analysis of political Twitter data. In: Mandal, J.K., Satapathy, S.C., Sanyal, M.K., Sarkar, P.P.,

Mukhopadhyay A. (eds.) Information Systems Design and Intelligent Applications, pp. 469–477. Springer, India (2015)

7. Duric, A., Song, F.: Feature selection for sentiment analysis based on content and syntax models. Decis. Support Syst. **53**(4), 704–711 (2012)

8. Cruz, N.P., Taboada, M., Mitkov, R.: A machine-learning approach to negation and speculation detection for sentiment analysis. J. Assoc. Inf. Sci. Technol., pp. n/a–n/a (2015)

9. Moraes, R., Valiati, J.F., GaviãoNeto, W.P.: Document-level sentiment classification: an empirical comparison between SVM and ANN. Expert Syst. Appl. **40**(2), 621–633 (2013)

10. Xia, R., Xu, F., Yu, J., Qi, Y., Cambria, E.: Polarity shift detection, elimination and ensemble: a three-stage model for document-level sentiment analysis. Inf. Process. Manag. (2015)

11. Liu, Y., Yu, X., Liu, B., Chen, Z.: Sentence-level sentiment analysis in the presence of modalities. In: Gelbukh, A. (ed.) Computational Linguistics and Intelligent Text Processing, pp. 1–16. Springer, Berlin Heidelberg (2014)

12. Peñalver-Martinez, I., Garcia-Sanchez, F., Valencia-Garcia, R., Rodríguez-García, M.Á., Moreno, V., Fraga, A., Sánchez-Cervantes, J.L.: Feature-based opinion mining through ontologies. Expert Syst. Appl. **41**(13), 5995–6008 (2014)

13. Esuli, A., Sebastiani, F.: SENTIWORDNET: a publicly available lexical resource for opinion mining. In: Proceedings of the 5th Conference on Language Resources and Evaluation (LREC'06), pp. 417–422 (2006)

14. Valitutti, R.: WordNet-Affect: an affective extension of WordNet. In: Proceedings of the 4th International Conference on Language Resources and Evaluation, pp. 1083–1086 (2004)

15. Cruz, F.L., Troyano, J.A., Pontes, B., Ortega, F.J.: ML-SentiCon: Un lexicón multilingüe de polaridades semánticas a nivel de lemas. Procesamiento del Lenguaje Natural **53**, 113–120 (2014)

16. Taboada, M., Brooke, J., Tofiloski, M., Voll, K., Stede, M.: Lexicon-based methods for sentiment analysis. Comput. Linguist. **37**(2), 267–307 (2011)

17. Ghosh, M., Animesh, K.: Unsupervised linguistic approach for sentiment classification from online reviews using SentiWordNet 3.0. Int. J. Eng. Res. Technol. **2**(9), (2013)

18. Perez-Rosas, V., Banea, C., Rada, M.: Learning sentiment Lexicons in Spanish. LREC (2012)

19. Clematide, S., Manfred, K.: Evaluation and extension of a polarity lexicon for German. In: Presented at the Proceedings of the First Workshop on Computational Approaches to Subjectivity and Sentiment Analysis, pp. 7–13 (2010)

20. Maks, I., Vossen, P.: Different approaches to automatic polarity annotation at synset level. In: Presented at the Proceedings of the First International Workshop on Lexical Resources, pp. 62–69 (2011)

21. Abdul-Mageed, M., Diab, M.: Toward building a large-scale Arabic sentiment lexicon. In: Presented at the Proceedings of the 6th International Global WordNet Conference, pp. 18–22 (2012)

22. Dehdarbehbahani, I., Shakery, A., Faili, H.: Semi-supervised word polarity identification in resource-lean languages. Neural Netw. **58**, 50–59 (2014)

23. Martín-Valdivia, M.-T., Martínez-Cámara, E., Perea-Ortega, J.-M., Ureña-López, L.A.: Sentiment polarity detection in Spanish reviews combining supervised and unsupervised approaches. Expert Syst. Appl. **40**(10), 3934–3942 (2013)

24. Balahur, A., Turchi, M.: Comparative experiments using supervised learning and machine translation for multilingual sentiment analysis. Comput. Speech Lang. **28**(1), 56–75 (2014)

25. Hsu, R., See, B., Wu, A.: Machine learning for sentiment analysis on the experience project. (2010)

26. Filho, P.P.B., Pardo, T.A., Alusio, S.M.: An evaluation of the brazilianportugueseliwc dictionary for sentiment analysis. In: Presented at the In 9th Brazilian Symposium in Information and Human Language Technology, Fortaleza, Ceara (2013)

27. Gonçalves, P., Araújo, M., Benevenuto, F., Cha, M.: Comparing and combining sentiment analysis methods. In: Proceedings of the First ACM Conference on Online Social Networks, New York, NY, USA, pp. 27–38 (2013)

28. Hutto, C.J., Gilbert, E.: Vader: a parsimonious rule-based model for sentiment analysis of social media text. In: Presented at the Eighth International AAAI Conference on Weblogs and Social Media (2014)
29. del P. Salas-Zárate, M., López-López, E., Valencia-García, R., Aussenac-Gilles, N., Almela, Á., Alor-Hernández, G.: A study on LIWC categories for opinion mining in Spanish reviews. J. Inf. Sci. **40**(6), 749–760 (2014)
30. Ye, Q., Zhang, Z., Law, R.: Sentiment classification of online reviews to travel destinations by supervised machine learning approaches. Expert Syst. Appl. **36**(3, Part 2), 6527–6535 (2009)
31. Sidorov, G., Miranda-Jiménez, S., Viveros-Jiménez, F., Gelbukh, A., Castro-Sánchez, N., Velásquez, F., Díaz-Rangel, I., Suárez-Guerra, S., Treviño, A., Gordon, J.: Empirical study of machine learning based approach for opinion mining in Tweets. In: Batyrshin, I., Mendoza, M.G. (eds.) Advances in Artificial Intelligence, pp. 1–14. Springer, Berlin Heidelberg (2013)
32. Pennebaker, J.W., Mayne, T.J., Francis, M.E.: Linguistic predictors of adaptive bereavement. J. Pers. Soc. Psychol. **72**(4), 863–871 (1997)
33. Francis, M.E., Pennebaker, J.W.: LIWC: linguistic inquiry and word count. Southern Methodist University, Dallas (1993)
34. Pennebaker, J.W., Francis, M.E., Booth, R.J.: Linguistic inquiry and word count, vol. 71. Lawrence Erlbaum Associates, Mahway (2001)
35. Ramírez-Esparza, N., Pennebaker, J.W., García, F. A., Suriá Martínez, R.: La psicología del uso de las palabras: un programa de computadora que analiza textos en español. Thepsychology of word use: a computerprogramthatanalyzestexts in Spanish, (2007)
36. Rushdi Saleh, M., Martín-Valdivia, M.T., Montejo-Ráez, A., Ureña-López, L.A.: Experiments with SVM to classify opinions in different domains. Expert Syst. Appl. **38**(12), 14799–14804 (2011)
37. Montejo-Ráez, A., Martínez-Cámara, E., Martín-Valdivia, M.T., Ureña-López, L.A.: Ranked WordNet graph for sentiment polarity classification in Twitter. Comput. Speech Lang. **28**(1), 93–107 (2014)
38. Chalothom, T., Ellman, J.: Simple approaches of sentiment analysis via ensemble learning. In: Kim, K.J. (ed.) Information Science and Applications, pp. 631–639. Springer, Berlin Heidelberg (2015)
39. Bouckaert, R.R., Frank, E., Hall, M.A., Holmes, G., Pfahringer, B., Reutemann, P., Witten, I.H.: WEKA—experiences with a Java open-source project. J. Mach. Learn. Res. **11**, 2533–2541 (2010)
40. Bhavsar, H., Ganatra, A.: A comparative study of training algorithms for supervised machine learning. Int. J. Soft Comput. Eng. **2**(4), 74–81 (2012)
41. Deng, N., Tian, Y., Zhang, C.: Support vector machines: optimization based theory, algorithms, and extensions. CRC Press, Boca Raton (2012)
42. Baldridge, J.: The opennlp project. openNLP. Available: https://opennlp.apache.org/ (2010). Accessed 18 May 2015
43. MacCartney, B.: Stanford classifer. The Stanford Natural Language Processing Group. Available http://nlp.stanford.edu/software/classifier.shtml. Accessed 18 May 2015
44. Anjaria, M., Guddeti, R.M.R.: Influence factor based opinion mining of Twitter data using supervised learning. In: 2014 Sixth International Conference on Communication Systems and Networks (COMSNETS), pp. 1–8 (2014)
45. Duyen, N.T., Bach, N.X., Phuong, T.M.: An empirical study on sentiment analysis for Vietnamese. In: 2014 International Conference on Advanced Technologies for Communications (ATC), pp. 309–314 (2014)
46. Chinthala, S., Mande, R., Manne, S., Vemuri, S.: Sentiment analysis on Twitter streaming data. In: Satapathy, S.C., Govardhan, A., Raju, K.S., Mandal, J.K. (eds) Emerging ICT for Bridging the Future—Proceedings of the 49th Annual Convention of the Computer Society of India (CSI), vol. 1, pp. 161–168. Springer, Berlin (2015)

Part II
Knowledge Acquisition and Representation

Chapter 5
Knowledge-Based System in an Affective and Intelligent Tutoring System

Ramón Zatarain Cabada, María Lucía Barrón Estrada and Yasmín Hernández Pérez

Abstract This book chapter presents an affective and intelligent tutoring system called Fermat that integrates emotion or affective states with an Intelligent Learning Environment. The system applies Knowledge Space Theory to implement the knowledge representation in the domain and student modules and Fuzzy Logic to implement a new knowledge tracing algorithm, which is used to track student's pedagogical and affective states. The Intelligent Learning Environment was implemented with two main components: an affective and intelligent tutoring system for elementary mathematics and an educational social network. The tutoring system generates math exercises by using a fuzzy system that is fed with cognitive and effective values. Emotion recognition was implemented by two methods: one for feature extraction of the face and one for feature classification using back-propagation neural networks. In addition to recognizing the emotional state of the user, our system gives emotional support through a pedagogical agent. Furthermore, an architecture of software is presented where the emotion recognizer collaborates with the affective and intelligent tutoring system inside a social network. Finally, we present a real-time evaluation with third-grade students in two different schools.

Keywords Affective computing · Intelligent tutoring systems · Social networks

R.Z. Cabada (✉) · M.L.B. Estrada
Departamento de Posgrado, Instituto Tecnológico de Culiacán, Culiacán,
Sinaloa, Mexico
e-mail: rzatarain@itculiacan.edu.mx

M.L.B. Estrada
e-mail: lbarron@itculiacan.edu.mx

Y.H. Pérez
Instituto Nacional de Electricidad y Energías Limpias, Tecnologías de la Información,
Cuernavaca, Mexico
e-mail: myhp@iie.org.mx

© Springer International Publishing AG 2017
G. Alor-Hernández and R. Valencia-García (eds.), *Current Trends on Knowledge-Based Systems*, Intelligent Systems Reference Library 120,
DOI 10.1007/978-3-319-51905-0_5

5.1   Introduction

In the last years, Intelligent Learning Environments (ILE) have incorporated the ability to recognize the student's affective state, in addition to traditional cognitive state recognition [1–3]. These learning systems and environments have special devices or sensors to measure or monitor facial actions, skin conductance, speech features, among others, and as result they recognize the emotional or affective state of the student [1, 4, 5]. Research works on affective computing include detecting and responding to affect. Affect detection systems identify frustration, interest, boredom, and other emotions [1, 4]. On the other hand, affect response systems transform negative emotional states (frustration, boredom, fear, etc.) to positive ones [5, 6].

If a software system that recognizes emotions needs to be implemented, then a small number of basic emotions are appropriate to be handled. Picard stated [7] "to date, there is no a validated theory or comprehensive work on emotions, to establish what are the emotions that influence the learning process". However, Ekman's work on facial expression analysis [8] described a subset of emotions including joy, anger, surprise, fear, disgust/contempt and interest, which have been used in new learning systems and include the recognition and management of emotions and/or feelings [1, 5, 9, 10].

In this work, we propose a framework that combines affective computing and social interaction in order to improve the learning process about math topics. We are using the official Mexican program in Mathematics for fourth grade students (http://issuu.com/sbasica/docs/ab-mate-3-baja1). Knowledge representation is implemented by using Knowledge Spaces Theory and Fuzzy Logic. For selecting the suitable response of the Intelligent Tutoring System, we used a decision network that selects the tutorial actions with the best expected effect on the student's affect and knowledge by using a dynamic decision network with a utility measure on both, learning and affect.

This book chapter is structured as follows: Sect. 5.2 presents related works in the field of Affective and Intelligent Learning Environments. Section 5.3 provides the architecture of the social network called Fermat and the affect-sensitive intelligent tutoring system. Section 5.4 presents the neural networks used to recognize emotional state and learning style of the student. Section 5.5 describes the decision network. An evaluation of the software Fermat with students is presented in Sect. 5.6 and conclusions and future work are discussed in Sect. 5.7. Finally, the concluding remarks and future directions are presented in Sect. 8.

5.2   Related Work

In recent years, an increase in the number of affective learning environments developed for different learning fields can be observed [11–13]. The abilities of these systems include detection and response to affective patterns. Affective detection systems observe and study the face, speech, conversation and other human features to detect frustration, interest, or boredom [14, 15]. Meanwhile, affective response systems handle and improve a student's negative emotion. There are seminal research works related to this problem [16, 17].

AutoTutor, an Intelligent Tutoring System (ITS) with conversational dialogues [10, 12, 18], is one of the most popular ITS. Affective AutoTutor was a new version which incorporates technologies from the field of Affective Computing. This new version works with students' affective and cognitive states.

The field of affective student modelling is also important [19]. In [20] the authors studied how emotions can occur during the learning process and what emotional states are optimal when a student is interacting in a learning environment. Burleson proposed a multimodal sensor system paying more attention to students' emotional experiences [21]. The emotion recognizer makes use of a pressure mouse, a wireless Bluetooth skin conductance sensor, a posture analysis seat, and a facial action unit analysis.

Nonetheless, most of these works have been focused on detecting emotions in a student by using special hardware sensors like posture chairs or skin conductance bracelets. There is a need to address this research work to most-common learning systems. The main contribution of this work is the creation of a new Affective and Intelligent Learning System for Mathematics where Knowledge representation is implemented in a more robust and natural form using knowledge spaces and fuzzy logic. Another contribution is that all this technology is integrated into Web-based learning platforms, not requiring special downloading and installation software, learning instructions for the user/student or specific hardware components.

In the field of ITS oriented to Mathematics there have been many developments, where a few of them use affect recognition and other learning strategies for increasing motivation like Gamification [22]. Lynnete is an ITS working in the Web for linear equations. It was designed for working in Android tablets and was implemented as a cognitive tutor [23]. Animal Watch is an ITS developed with two goals: Teaching with efficacy arithmetic and pre-algebra and increasing the confidence and liking of mathematics in young students. The system integrates mathematics with science, art, and technology. MathTutor [24] is a Web-based ITS designed to help 6th grade, and 1st and 2nd year high school students, to learn Mathematics by performing exercises (learning by doing). The tutoring system includes a large number of exercises and problems and a complete learning management system that enables teachers to create class lists, assign jobs to a group or to an individual student, and view reports on student progress. Wayang Outpost Tutor [11] is an adaptive math tutoring system that helps students learn to solve standardized-test questions. The system uses multimedia to help students solve problems in Geometry.

Next, we present a comparative analysis of this proposal with other influential intelligent tutoring systems for mathematics (Table 5.1).

We can observe in Table 5.1 that there are different knowledge representations for the tutoring systems but production rules and bayesian networks are the most common. In Fermat, the knowledge representation of the domain and student model

Table 5.1 Comparison of Fermat with other similar intelligent tutoring systems

ITS	Knowledge representation	Affect recognition/handling	Area
Lynnete	Production rules (ACT-R)	No	Linear equations
Animal watch	Reinforcement learning and Bayesian networks	No	Arithmetic and pre-algebra
Wayang outpost	Bayesian networks	Yes	Geometry
Math tutor	Production rules (ACT-R)	No	Mathematics (elementary and high school)
Fermat	Knowledge space and fuzzy sets	Yes	Arithmetic

was implemented with Knowledge spaces and Fuzzy Logic. The main advantage of the first method is that the student knowledge can be seen as a subset (sub-tree implemented) of all knowledge possessed by the expert in the domain (module) and that makes it easy to integrate and work with both models (expert and students). With respect to Fuzzy Logic, it is a technique that reflects how people think. It models our way of thinking, of communicating with each other, and our reasoning. Both methods make the knowledge representation of Fermat more robust and natural.

5.3 Fermat Architecture and ITS Knowledge Representation

The Fermat Educational Social Network has the basic and common functionalities of all social networks, but its main feature is the inclusion of an ITS that offers to users course content in a customized style. Fermat's members are associated with personal, academic and affective information in static and dynamic profiles, which are gathered during the user navigation process on Fermat. The static profile contains initial information of the user (e.g. personal and academic information) while the dynamic profile stores the information generated during a session with the intelligent tutor (e.g. emotional states, time using the tutor, lessons covered, etc.).

5.3.1 The System Archetypes

The archetypes that make up the social network and form the most stable part of the system are the following (Fig. 5.1 shows the relationships among the archetypes):

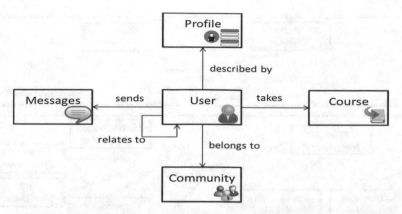

Fig. 5.1 System archetypes. The relationships among the different archetypes are depicted. The archetype users is an abstraction of the entire set of users of the social network

- Users represent an abstraction of the entire set of users that make up the social network.
- Profile is a unit of information that contains personal data.
- Community represents a group of users who have something in common, such as courses and teachers, to mention but a few.
- Message is an abstraction that describes the device that allows communication among users.
- Courses are a set of ITS that users can access to enhance their education.

5.3.2 Fermat's Architectural Style

Once the components of the social network have been established, the Fermat's three-layer architecture is presented. This layered architectural design allows abstracting system components at different levels in order to develop each one of them in an independent way. This layered design allows scalability and easy maintenance because its tasks and responsibilities are distributed. The layers are: Presentation, Business, and Data Access. This architectural style was proposed due to its support to add new features, so future changes to the social network can be done without affecting other components. It also enables a better control over data persistence. Each layer of the architecture has a function and it is shown in Fig. 5.2.

The *presentation layer* directly interacts with the user, so it was designed to be user-friendly. This layer only communicates with the Business layer and contains a set of graphical user interfaces both for the Web-based and mobile versions of Fermat.

The *business layer* contains all components required for a user to be able to access the social network services. This layer acts as a brokering service between the user and the data. In this layer, the main features/functionalities of the social

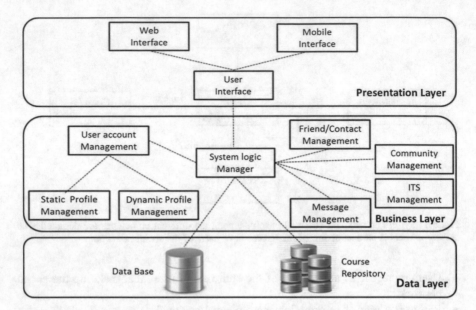

Fig. 5.2 Fermat's general architecture. The architecture is composed by three layers: data layer, business layer, and presentation layer

network are implemented (e.g. user accounts, static and dynamic user profiles, user contacts, user communication via messages, user communities and the intelligent tutoring system that is the most important component of this layer).

The *data access layer* contains the database that stores the information of the members of the social network. Furthermore, this layer stores educational content into a repository of courses. Additionally, it contains all the configuration tables allowing the operation of the modules and services offered by Fermat. A database management system is located in this layer which maintains the data persistence by executing CRUD (Create, Read, Update and Delete) operations. Users do not have direct access to this layer.

5.3.3 Intelligent Tutoring System into Fermat Social Network

The integration scheme between the Learning Social Network Fermat and the Intelligent Tutoring System is shown in Fig. 5.3. Social network users (students) are associated with personal, academic and affective information in a profile, which is obtained from the user navigation process. There is a neural network to obtain the affective state of the students which is used by the tutoring module. Besides the neural network, the emotion recognizer uses the web camera to capture important

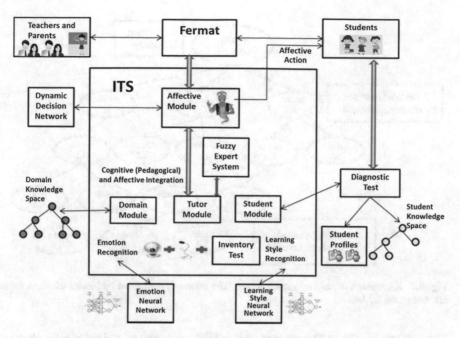

Fig. 5.3 Intelligent tutoring system into Fermat social network. The architecture of the ITS is depicted and its integration with Fermat social network

features from the student face. For future work, an Inventory Learning Style Questionnaire (ILSQ) and a neural network will be used for learning style recognition.

5.3.4 Knowledge Representation in the Intelligent Tutoring System

The domain and the student modules are a tree structure which applies some of the concepts related to Knowledge Space Theory [25] that provides a sound foundation for structuring and representing the knowledge domain of the ITS. A course can be seen as a discipline-specific knowledge space (a particular tree diagram) containing blocks, which in turn are made by lessons. The total of nodes in the tree represents the domain or expert knowledge. Figure 5.4 shows the knowledge domain of the math course with subjects related to arithmetic operations such as multiplication and division and topics like fractions.

The domain models are quantitative representations of expert knowledge in a specific domain; therefore we try to define a knowledge representation that matches the structure shown in third grade books for the official program in Mexico.

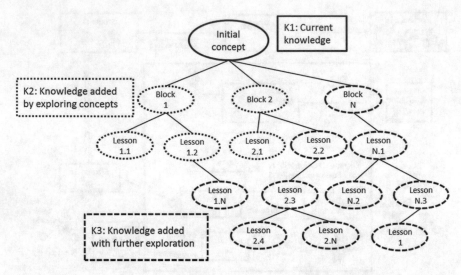

Fig. 5.4 Knowledge domain of a math course. The course is composed by blocks which in turn are composed by lessons

The Knowledge Space Theory uses concepts of combinatory and probability theory to model and empirically describe particular areas of knowledge (the knowledge of a particular student). In formal terms, the theory indicates that a knowledge structure is a pair (Q, K) in which Q is a nonempty set, and K is a family of subsets of Q, which contains at least one and one empty set $\emptyset$. The set Q is called the domain of knowledge structure and its elements are called questions or items and the subsets in the K family are labeled (knowledge) as states. Occasionally it may be said that K is a knowledge structure on a set Q in the sense that (Q, K) is a knowledge structure. The specification of the domain can be omitted without ambiguity since we have $\cup K = Q$. For example, we can represent the model of knowledge as:

$$Knowledge = \{\emptyset, \{Block1, Lesson1.1, Lesson1.2\}, \{Block2, Lesson2.1, Lesson2.2\}\}$$

The student module provides information about student knowledge and learning aptitudes. The module identifies what the student's knowledge is through a diagnostic test. The student knowledge can be seen as a subset (sub-tree implemented) of all knowledge possessed by the expert in the domain (module) and this is stored in a student profile, as shown in the right side of Fig. 5.3. For the tutoring module (recall Fig. 5.3 again), a fuzzy expert system was implemented with a new knowledge tracing algorithm, which is used to track student's pedagogical states, applying a set of rules. The fuzzy system uses input linguistic variables such as error, help, and time. These variables are loaded while the student solves an exercise. The output variable of the fuzzy system is the difficulty of the next exercise. The proposed fuzzy sets, for each linguistic variable are:

Table 5.2 A sample of fuzzy rules of the expert system

No.	Rule
Rule 1	If (Error is low) and (Help is little) and (Time is very fast) then (Difficulty is very_hard)
Rule 2	If (Error is low) and (Help is little) and (Time is fast) then (Difficulty is very_hard)
Rule 3	If (Error is low) and (Help is little) and (Time is normal) then (Difficulty is very_hard)
Rule 4	If (Error is low) and (Help is little) and (Time is slow) then (Difficulty is hard)
Rule 5	If (Error is low) and (Help is little) and (Time is very_slow) then (Difficulty is hard)
Rule 6	If (Error is low) and (Help is normal) and (Time is slow) then (Difficulty is basic)
…	
Rule 41	If (Error is many) and (Help is helpful) and (Time is very_slow) then (Difficulty is very_easy)

Error = {low, normal, many}
Help = {little, normal, helpful}
Time = {very fast, fast, slow, very slow}
Difficulty = {very easy, easy, basic, hard, very hard}.

One important step of our fuzzy expert system is to evaluate the fuzzy values of the input variables. Table 5.2 shows a sample of some of the fuzzy rules that are used in the system.

In order to evaluate the conjunction of the rule antecedent, we applied the Eq. 5.1:

$$\mu_{A \cap B \cap C \ldots \cap Z}(x) = \min \left[\mu_A(x), \mu_B(x), \mu_c(x), \ldots, \mu_Z(x) \right] \qquad (5.1)$$

To evaluate disjunction, we applied the Eq. 5.2:

$$\mu_{A \cup B \cup C \ldots \cup Z}(x) = \max \left[\mu_A(x), \mu_B(x), \mu_c(x), \ldots, \mu_Z(x) \right] \qquad (5.2)$$

For instance, to evaluate the next fuzzy rule:

IF   Error is low (0.3)
AND   Help is little (0.2)
AND   Time is very-fast (0.1)
THEN   Difficulty is very-hard (0.1).

Equation 5.1 is applied which results Eq. 5.3:

$$\mu_{very_{hard}}(Difficulty) = \min \left[\mu_{low}(Error), \mu_{little}(Help), \mu_{very_{fast}}(time) \right]$$
$$= \min [0.3, 0.2, 01] = 0.1 \qquad (5.3)$$

Next, we present how the student solves the exercises with the help and support of the intelligent tutoring system. The basic structure of an XML-based file consists of an array of objects which contains the *divisor* and *dividend* attributes which are shown to the student. The *quotient*, *reminder* and *mul* attributes contain the correct answers.

An initial exercise is presented to the student through a graphical interface; students can enter answers they think are correct, while the intelligent tutor dynamically checks the corresponding XML-based file to verify the answer and to provide responses to them. The initial exercise has a difficulty level that was set for each student profile according the result in the diagnostic test completed by the student. The difficulty level of the next exercises can be modified depending on the student's performance in solving each mathematic exercise.

The functionality of how responses are evaluated and the path considered by the solution process are shown in Fig. 5.5. In this context, the premise is simple. The ITS waits for an entry value t, and it verifies that the value is correct. When a correct value is entered, the ITS moves to the next box; then it will wait for the next input value. Otherwise, the ITS sends a message through a pedagogical agent about the type of error found in the answer and then it waits for a student response. This action is repeated until the division operation is finished.

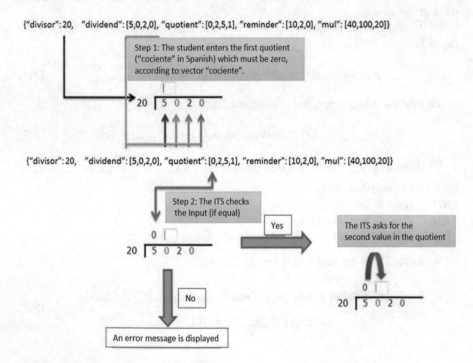

Fig. 5.5 Evaluation of an example of the division arithmetic operation. Every step in the operation is explained

5.4   Recognizing Emotional States

The main goal of the artificial neural networks is to recognize the emotional state of the student. At the beginning of a course, the student's knowledge state is evaluated and calculated by using a diagnostic test. Later, a student interacts with the ITS and a neural network infers, in real-time, the emotional state of the student based on facial expressions.

5.4.1   Artificial Neural Network for Recognizing Emotional States

The method used for the detection of visual emotions is based on Ekman's theory. The recognition system was built in three stages: the first stage was an implementation to extract features from facial expressions in a corpus used to train the neural network. The second one consisted of the implementation of a back-propagation neural network. The third stage integrated extraction and recognition into a client-server model. In this way, the features extraction process is carried out on the client side and the emotion recognition process is located on the server side. Figure 5.6 shows the emotion recognition system.

Feature extraction and emotion recognition were implemented by using OpenCV, JavaCV, and Weka libraries (the back-propagation neural network classifier was built from the source code of Weka). The layered neural network is formed by the following items: (1) an input layer of ten source neurons which read ten feature values corresponding to distances extracted from each face. Six distances are obtained from the mouth, two distances from the left eye and two from the right eye; (2) a hidden layer with 10 sigmoid neurons; and (3) an output layer of 7 linear neurons corresponding to the 7 emotions the network helps to recognize. Other parameters for training the neural network were the learning rate (0.3) and the epoch number (2000).

5.4.1.1   Training and Testing the Network with Emotional States

For training and using the neural network, we used the corpus RAFD (Radboud Faces Database) [26] which is a database with 8040 different facial expressions containing a set of 67 models including men and women. With this database, 96% of correct recognitions were obtained. We developed a neural network in Matlab with sigmoid hidden neurons and linear output neurons. The network was trained with the Levenberg-Marquardt back-propagation algorithm. Regression Values that measure the correlation between outputs and targets had values very close to 1, meaning an almost perfect lineal association between target and actual output

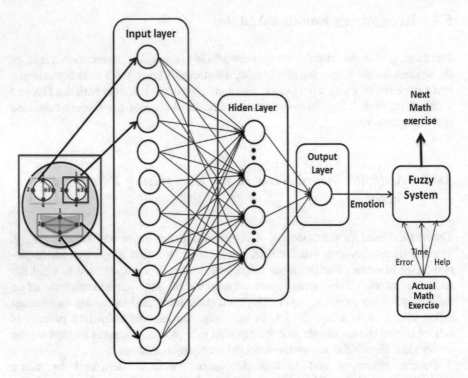

Fig. 5.6 Emotion recognition system. The system is composed by a back propagation neural network connected with the fuzzy system

values (independent and dependent variables). We also obtained excellent results with the Weka tool with a success rate of 96.9466% in the emotion recognition process.

5.5 The Decision Network

The tutorial actions are composed by an affective and a pedagogical component. The affective component of a tutorial action tries to promote a student into a positive affective state and the pedagogical component aims to convey knowledge. The decision process is represented as a dynamic decision network (DDN) as is shown in Fig. 5.7. The DDN included in the model is used to predict how the available tutorial actions can influence the student's knowledge and the student's affect state, given her/his current state.

Our model uses multi-attribute utility theory to define the necessary utilities [27, 28]. From this perspective, the DDN establishes the tutorial action considering two utility measures: one on learning and one on affective aspects, which are combined to obtain the global utility by a weighted linear combination. These utility functions

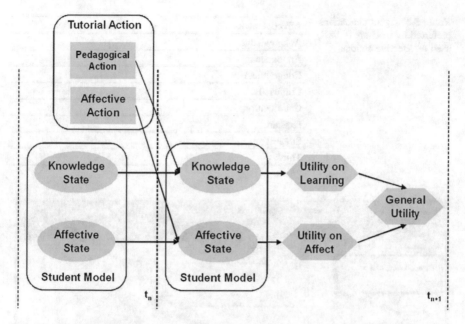

Fig. 5.7 High level DDN for the affective tutor model. The network is composed by two time slices, the current and the predicted one with base on current states

are the means that allow educators adopting the system to express their preferences towards learning and affective factors.

The utility for learning is measured in terms of how much the student's knowledge is improved by the tutorial action given her/his current knowledge. Similarly, the utility for affect is measured in terms of how much the student affect improves as a result of the action. Finally, the overall utility is computed as a weighted sum of these two utilities. Thus, the intelligent tutor calculates the utility for each tutorial action considering the current state, and it selects the tutorial action with the maximum expected utility.

The affective actions are shown in Table 5.3. These actions are the results of a study conducted to evaluate the expressivity of a pedagogical agent. In the study, 20 teachers were asked to select appropriate affective actions to be presented according to several tutorial scenarios.

5.5.1 Integrating Affect into the ITS

As we mentioned before, the utility on learning represented by the level reached in the difficulty of the exercises, is combined with the utility of affection, which will be used by the pedagogical agent to respond in a better way to negative affective states. For instance, if a student reaches a level of difficulty with fuzzy value *very*

Table 5.3 Agent animations preferred by teachers to be used as affective actions

Affective action
Acknowledge
Announce
Congratulate
Confused
Get attention
Explain
Suggest
Think

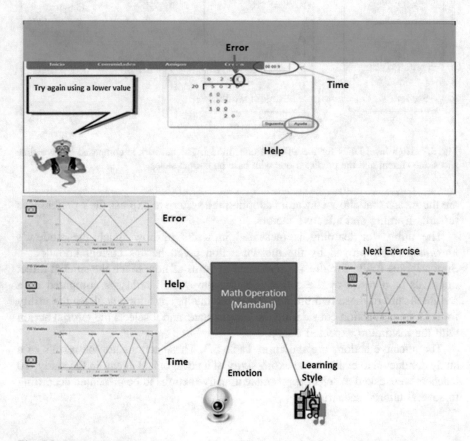

Fig. 5.8 Input and output fuzzy variables. The output variable of the fuzzy system, in combination with emotion and learning style, determines the next exercise

easy, and an emotional state of sadness, then the ITS has to respond accordingly (adjusting the level of exercise and emotional state). For every student there is a static and a dynamic profile, which store particular and academic information like affective states and scoring results. In the ITS, a fuzzy expert system was implemented with a new knowledge tracing algorithm, which is used to track student's pedagogical states, applying a set of rules.

As we established before (Sect. 5.3.3), our fuzzy system uses input linguistic variables error, help, and time, but now adding new fuzzy variables named Emotion, and Learning Style (this last variable for future implementation) (Fig. 5.8). These variables are loaded when the student solves an exercise. The output variable of the fuzzy system determines the difficulty and type of the next exercise. The type will be defined according to the learning style assigned to the student.

5.6 Evaluation and Results

In order to validate the legitimacy of the ITS inside the social network, a real-time evaluation was made with third-grade students in two different schools. We applied pre-test, tutor intervention, and post-test to students before and after they learned integer multiplication and division operations with Fermat. This method of evaluation or comparison in intelligent tutoring systems is fully documented in [29]. The evaluation experiment utilized 33 students (males and females) ages 8 and 9 years old, from public and private schools.

5.6.1 Fermat Real-Time Evaluation

Fermat (Social network and affective tutoring system) was also evaluated with third-grade students in two different schools (public school *Benito Juárez* and private school *Instituto Chapultepec*) in the city of Culiacán, México (Fig. 5.9). The method of the evaluation consisted of a brief introduction of the tool (Fermat), registration into the social network of Fermat, a pre-test evaluation, learning integer multiplications and division using the affective tutoring system of Fermat, and a post-test evaluation with a sample of 33 students and with the help of teachers.

The *introduction* to Fermat was carried out in 15 min with a very positive attitude from the students, where the main issue of the students was to compare Fermat with Facebook.

We had some problems in the *registration into Fermat* because the students have no familiarization with respect to functionalities of the tool like capturing personal data into the profile or sending friendship messages to their classmates. The activity was carried out in 30 min.

The *pre-test evaluation* took about 15 min, and at the end the students received their final scores (0–10 scale). Some of the students were very enthusiastic with the system giving the results in an attractive interface.

After finishing the diagnostic test, in the *Multiplication and Division with Affective Tutor* activity, the students received a brief introduction of the tutoring

Fig. 5.9 Real-Time evaluation of Fermat. Thirty-three students from a public and a private school participated in the evaluation

Fig. 5.10 Comparison of pre-test and post-test results. The student answered similar test containing 10 exercises

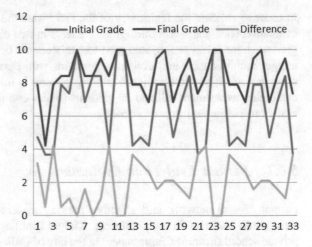

system. They received information about answer formats, operation procedures, and about the pedagogical agent or avatar (Genius). The activity was developed in about 45 min for each arithmetic operation, where the students solved several exercises with different levels of complexity.

In the *post-test evaluation*, a test containing 10 computer exercises was gave to students, and they had 15 min to solve them. Figure 5.10 illustrates the pre-test and post-test results obtained by 33 students (9 from a public school and 24 from a private school).

As we can observe, the initial and final grades show 27 students getting better scores or improvement, and show 6 students having the same score (no improvement). A detailed analysis of the results finds interesting facts. One of them is that, in the pre-test, half of the students in the private school had a grade lower than 6 and a third of the students in the public school had it. On the other hand, in the post-test, all of the students in the private school had grades higher than 6 and most students (more than 90%) of the public school had it as well. According to teachers that were consulted after we obtained the results, they believe that the procedure (algorithm) where the children solve the problems, the ITS ability to recognize the student's affective and pedagogical state, and the help from the pedagogical agent give many benefits compared to traditional classroom teaching. They also mentioned that students that did not improve their score were students in the top 10% of the class, and students having greater improvements were students in the bottom 10% of the class. From these observations we believe that we have to present more challenging problems for top students.

5.7 Conclusions and Future Directions

In this work, we have proposed a framework which includes affect, learning styles and social interaction in order to improve the learning of math in children. We have described the components of the framework including the implementation of the knowledge system of the domain and student modules. Both modules were implemented by using knowledge spaces and fuzzy logic. We also described an affect recognizer implemented with a neural network. This system identifies the student emotion which is used, together with some cognitive values, by the fuzzy system. The results up to now are encouraging. We are implementing our own recognizers because we need to use them in a Web platform where the social network can be accessed from different computers like laptops or Mobile smartphones.

As future work, we are conducting a controlled user study and in this way, trying to confirm our hypothesis: the learning process is improved when the affective state, learning styles and social interaction are considered besides the knowledge. We are also concluding the implementation of the learning style module and the integration of this module to the fuzzy system. This Module consists of a predictive engine used to dynamically identify the student´s learning style whenever the tutoring system is running. At the beginning an interpreter will select content (learning lessons) based upon the student's learning style obtained from the student profile. The learning style of a student can be modified according to cognitive and affective evaluations applied to the student. Last, we are working to incorporate emotion recognition using a multimodal approach that combines speech, facial, and brainwaves detection.

Another research line we are working with Fermat is in the area of educational applications for TV, Cloud-based platforms for Learning, and Learning Management Systems with SCORM Metadata [30–33].

The intelligent tutoring system and recognizers were implemented by using different software tools and programming languages. The presentation layer of the social network was implemented with CCS3, HTML 5 and Javascript. For the business layer (the intelligent tutor mainly) we use Java and JSP. For the data layer we use XML and MySQL.

Acknowledgments The work described in this book chapter is fully supported by a grant from the DGEST (Dirección General de Educación Superior Tecnológica) in México under the program *Support to projects of scientific and applied research.*

References

1. Jaques, P.A., Vicari, R.M.: A BDI approach to infer student's emotions in an intelligent learning environment. Comput. Educ. **49**(2), 360–384 (2007)
2. Baker, R.S., D'Mello, S.K., Rodrigo, M.M.T., et al.: Better to be frustrated than bored: the incidence, persistence, and impact of learners' cognitive–affective states during interactions with three different computer-based learning environments. Int. J. Hum. Comput Stud. **68**(4), 223–241 (2010)
3. Desmarais, M.C., Baker, R.S.J.D.: A review of recent advances in learner and skill modeling in intelligent learning environments. User Model. User-Adap. Inter. **22**(1–2), 9–38 (2012)
4. Conati, C., Maclaren, H.: Empirically building and evaluating a probabilistic model of user affect. User Model. User-Adap. Inter. **19**(3), 267–303 (2009)
5. D'Mello, S.K., Picard, R.W., Graesser, A.C.: Towards an affective-sensitive AutoTutor. IEEE Intell. Syst. Spec. Issue Intell. Educ. Syst. **22**(4), 53–61 (2007)
6. Du Boulay, B.: Towards a motivationally intelligent pedagogy: how should an intelligent tutor respond to the unmotivated or the demotivated? In: Calvo, R.A., D'Mello, S.K. (eds.) New Perspectives on Affect and Learning Technologies, pp. 41–52. Springer, New York (2011)
7. Picard, R.W., Papert, S., Bender, W., et al.: Affective learning—a manifesto. BT Technol. J. **22**(4), 253–269 (2004)
8. Ekman, P., Friesen, W.: Unmasking the Face: a Guide to Recognizing Emotions from Facial Clues. Prentice-Hall, Englewood Cliffs, NJ (1975)
9. Forbes-Riley, K., Litman, D.J.: Adapting to student uncertainty improves tutoring dialogues. In: Proceedings of International Conference on Artificial Intelligence in Education, pp. 33–40 (2009)
10. D'Mello, S., Graesser, A.: AutoTutor and affective AutoTutor: learning by talking with cognitively and emotionally intelligent computers that talk back. ACM Trans. Interact. Intell. Syst. **2**(4), 1–39 (2012)
11. Arroyo, I., Woolf, B.P., Burleson, W., et al.: A multimedia adaptive tutoring system for mathematics that addresses cognition, metacognition and affect. Int. J. Artif. Intell. Educ. **24**(4), 387–426 (2014)
12. Nye, B.D., Graesser, A.C., Hu, X.: AutoTutor and family: a review of 17 years of natural language tutoring. Int. J. Artif. Intell. Educ. **24**(4), 427–469 (2014)
13. Sabourin, J.L., Lester, J.C.: Affect and engagement in game-based learning environments. IEEE Trans. Affect. Comput. **5**(1), 45–56 (2014)
14. Bosch, N., Chen, Y., D'Mello, S.: It's written on your face: detecting affective states from facial expressions while learning computer programming. In: Proceedings of International Conference on Intelligent Tutoring Systems, pp. 39–44 (2014)
15. Bosch, N., D'Mello, S., Baker, R., et al.: Automatic detection of learning-centered affective states in the wild. In: Proceedings of the 20th International Conference on Intelligent User Interfaces, pp. 379–388 (2015)
16. D'Mello, S., Olney, A., Williams, C., et al.: Gaze tutor: a gaze-reactive intelligent tutoring system. Int. J. Hum. Comput Stud. **70**(5), 377–398 (2012)
17. Harley, J.M., Trevors, G.J., Azevedo, R.: Clustering and profiling students according to their interactions with an intelligent tutoring system fostering self-regulated learning. J. Educ. Data Min. **5**(1), 104–146 (2013)

18. Graesser, A.C., D'Mello, S.K., Strain, A.C.: Emotions in advanced learning technologies. In: International Handbook of Emotions in Education, pp. 473–493 (2014)
19. Chrysafiadi, K., Virvou, M.: Student modeling approaches: a literature review for the last decade. Expert Syst. Appl. **40**(11), 4715–4729 (2013)
20. Lester, J.C., McQuiggan, S.W., Sabourin, J.L.: Affect recognition and expression in narrative-centered learning environments. In: Calvo, R.A., D'Mello, S.K. (eds.) New Perspectives on Affect and Learning Technologies, pp. 85–96. Springer, New York (2011)
21. Burleson, W.: Advancing a multimodal real-time affective sensing research platform. In: Calvo, R.A., D'Mello, S.K. (eds.) New Perspectives on Affect and Learning Technologies, pp. 97–112. Springer, New York (2011)
22. Kapp, K.M.: The gamification of learning and instruction: game-based methods and strategies for training and education. Pfeiffer, San Francisco (2012)
23. Long, Y., Aleven, V.: Gamification of joint student/system control over problem selection in a linear equation tutor. In: Proceedings of International Conference on Intelligent Tutoring Systems, pp. 378–387 (2014)
24. Aleven, V., McLaren, B.M., Sewall, J.: Scaling up programming by demonstration for intelligent tutoring systems development: an open-access web site for middle school mathematics learning. IEEE Trans. Learn. Technol. **2**(2), 64–78 (2009)
25. Doignon, J.P., Falmagne, J.C.: Knowledge spaces. Springer, New York (1999)
26. Langner, O., Dotsch, R., Bijlstra, G., et al.: Presentation and validation of the Radboud Faces Database. Cogn. Emot. **24**(8), 1377–1388 (2010)
27. Clemen, R.T.: Making Hard Decisions. Duxbury Press, Belmont CA (2000)
28. Murray, R.C., VanLehn, K.: DT tutor: a decision-theoretic dynamic approach for optimal selection of tutorial actions. In: Proceedings of International Conference on Intelligent Tutoring Systems, pp. 153–162 (2000)
29. Woolf, B.P.: Building Intelligent Interactive Tutors: Student-Centered Strategies for Revolutionizing e-Learning. Morgan Kaufmann, Burlington, MA (2010)
30. Vásquez-Ramírez, R., Alor-Hernández, G., Sánchez-Ramírez, C., et al.: AthenaTV: an authoring tool of educational applications for TV using android-based interface design patterns. New Rev. Hypermedia Multimedia **20**(3), 251–280 (2014)
31. Vásquez-Ramírez, R., Alor-Hernández, G., Rodríguez-González, A.: Athena: a hybrid management system for multi-device educational content. Comput. Appl. Eng. Educ. **22**(4), 750–763 (2014)
32. Vásquez-Ramírez, R., Bustos-Lopez, M., Montes, A.J.H., et al.: An open cloud-based platform for multi-device educational software generation. In: Proceedings of the 4th International Conference on Software Process Improvement, pp. 249–258 (2016)
33. Esteban-Gil, A., Fernández-Breis, J.T., Castellanos-Nieves, D., et al.: Semantic enrichment of SCORM metadata for efficient management of educative contents. Procedia-Social Behav. Sci. **1**(1), 927–932 (2009)

Chapter 6
A Software Strategy for Knowledge Transfer in a Pharmaceutical Distribution Company

Mario Barcelo-Valenzuela, Patricia Shihemy Carrillo-Villafaña, Alonso Perez-Soltero and Gerardo Sanchez-Schmitz

Abstract This work solves knowledge transfer problems faced by a family owned pharmaceutical distribution company. The main objectives of this project are to improve knowledge transfer efficiency, recover outdated knowledge and improve the company's operation. As years transpire, explicit knowledge can become obsolete and inaccessible for several reasons, i.e. staff's retirement or upgrades in the software being used. In the company's case, explicit knowledge was not identified and was saved in different software versions and hardware, leading to delays and rework attributed to knowledge's inaccessibility. A software strategy was proposed and implemented to solve this problematic. Favorable results were achieved. Knowledge transfer between departments was improved, knowledge from older versions or software were recovered, knowledge was centralized and a middleware was successfully implemented in one of the departments. Company personnel can locate, access and implement knowledge and the company's operations has improved.

Keywords Knowledge · Knowledge transfer · Software strategy · Knowledge recovery

6.1 Introduction

In recent times, knowledge has emerged as an important concept in commercial debates. Organizations now recognize it as a valuable strategic resource, and it requires a coherent management that is based on the company's established strategy [1]. This task is the joint effort between operation personnel and knowledge management personnel. Burmeister and Deller [2] argue the necessity of identifying organizational practices that support knowledge transfer for their strategic implementation of knowledge management in an organization's activities.

M. Barcelo-Valenzuela (✉) · P.S. Carrillo-Villafaña · A. Perez-Soltero · G. Sanchez-Schmitz
Universidad de Sonora, Hermosillo, Sonora, Mexico
e-mail: mbarcelo@industrial.uson.mx

© Springer International Publishing AG 2017
G. Alor-Hernández and R. Valencia-García (eds.), *Current Trends on Knowledge-Based Systems*, Intelligent Systems Reference Library 120,
DOI 10.1007/978-3-319-51905-0_6

Since knowledge is substantially collaborative, it requires the use of diverse technologies as a support tool for its management [3]. In professional service oriented companies, survival depends on the exploitation of employee skills and knowledge [4]; and it's shared and created through Knowledge Management Systems (KMS).

There are several studies on KMS in the literature, such as those belonging to. Others, like [8], are intended to facilitate the selection and development of KMS methodologies while revealing strengths and weaknesses in the developmental area of KMS, which exist because most methodologies encompass the identification, assessment and classification of organizational knowledge and only consider success relevant in the short term. Knowledge Management (KM) has focused on creating methods for the selection and use of success factors; e.g. [9], has proposed a framework that identifies the most important KMSs that can achieve an organization's objectives. Through the study of different KM models, Lee and Wong [10] state that the differences between systems destined for large enterprises versus small-medium enterprises (SMEs), reside in the indicators used that are based on owner-administrator knowledge in SMEs. Their knowledge is considered the main source of knowledge in SMEs but has seldom been included as an indicator for large enterprises.

Models and proposals were identified, e.g. [5] seeks to maximize the efficiency of a company that shares knowledge. Some authors, e.g. [7, 11, 12], include several processes, however, knowledge transfer is not sufficiently developed. According to [13], the process of knowledge transfer, specifically its consequences, has not been amply studied. In addition, some studies of knowledge transfer highlight the finding of barriers and social and cultural factors that affect the transfer. These studies, i.e. are important references, however, they do not answer the issue raised in this study [14–16]. According to [17] company innovation is encouraged when knowledge transfer is proactive, this is achieved through new technologies that encourage knowledge transfer between users.

Although there are several studies of KM, and KMS in particular, knowledge transfer is discussed in different perspectives, such as the collaboration, transmission, or sharing of knowledge between a transmitter and a receiver. However, little importance has been given to issues during the transfer. Explicit knowledge is possessed by an individual, that knows where it is located (electronic or physically) and how to use it, however, co-workers might be unaware they have this information, where it is located and might not have access to it. This knowledge can be stored in a computer device or be a printed document, and be considered obsolete or rarely used; or can be stored in an incompatible software format. The objective of this study is the development of a software strategy for the transfer of knowledge, stored in different software versions and hardware types, for the improvement of a company's operation.

The document with a theoretical background of concepts related to knowledge, its context in knowledge management and the transfer of knowledge. Then, a model and its framework are proposed. This section includes a brief explanation on common issues found during knowledge transfer and the proposed model which consists of 4 stages: identification, capture/storage, transfer/visualization and application. An explanation of each stage is included to enable its application in other organizational environments. The section that follows presents a general description of the problems a pharmaceutic distribution company faces where the framework was applied. It also includes specifics on the implementation of each stage, and their evaluations. Finally, the conclusions illustrate the results attained and the benefits achieved with the software strategy that was developed for the company.

6.2   Theoretical Background

6.2.1   Knowledge and Knowledge Management

Knowledge might be the most important competitive advantage source in this turbulent global economy [18]. The word "knowledge" has several interpretations. Churchman [19] had defined knowledge as the process by which data lead to information; which first was analyzed and then communicated, leading to knowledge. One of the most recognized definitions of knowledge is that it is a belief that is true and justified [20]. Creating new knowledge requires an organization and for teams to be involved in a non-stop process of personal and organizational self-renewal [21]. Hedlund [22] expresses that knowledge can be classified as tacit and explicit. The first is based on personal experience and the latter is more precise, formal and documented. Tacit knowledge is more difficult to identify, evaluate, and absorb because it is embedded in organizational practices and informal rules, routines and processes [23]. It is transferred mainly through observation and face-to-face interactions [24, 25]. Individuals that possess this type of knowledge "know more than they can tell" and are often unaware that they have it [25].

According to [6], KM theory has evolved on practical interest in managing knowledge for an organization's benefit rather than on a universal understanding of knowledge. They concluded that advanced information technologies (e.g. the internet, intranets, extranets, browsers, data warehouses, data mining techniques, and software agents) can be used to systematize, enhance, and expedite large scale intra- and extra-firm knowledge management. KM can be seen as the organizational processes that seek the synergistic combination of data and information, with the processing capacity of information technologies and the creative and innovative capacity of human beings [26]. Another approach to KM is as the device by which managers can generate, communicate, and exploit knowledge (usable ideas) for personal and organizational benefit [27].

KM is broad, multi-dimensional and covers most aspects of an enterprise's activities. Enterprises need to set broad priorities and integrate the goals of managing intellectual capital with their corresponding effective knowledge processes. According to [28], KM is the systematic, explicit and deliberate administration and operation of intellectual capital and knowledge of processes, and it is based on people, technologies and/or resource management. For [29], KM is the process of identification, documentation, organization and dispersal of intellectual assets that are critical for organizational performance in the long term. To promote KM, a collaborative climate where employees support and help each other has to be fostered instead of a competitive climate [30]. Many studies have suggested that management incentive plays a major role in the success of KM initiatives [31–33].

Tacit knowledge resides internally in a person and explicit knowledge can be modeled and can be perceived through possible representations using modeling, expressions, and/or by other means [34]. Explicit knowledge can be found in documents, data bases, information technologies, patents, licenses, scientific equations, specifications, instructions, manuals, amongst other [35]; it is knowledge that is easy to share, express and store. On the other hand, tacit knowledge is highly personal and deep-rooted in the human mind, it is hard to separate from individuals which possess it and can be hard or impossible to code and transfer [36]. Knowledge derived from personal experience and intuition is also hard to understand by other individuals [37]. Due to these facts, tacit knowledge has a significant influence in the knowledge transfer process.

6.2.2   Knowledge Transfer

Knowledge transfer refers to the process of communicating knowledge from one agent to another [38]. Krishnaveni and Sujatha [39] define it as the cession of knowledge from one place, individual or composition to another. In an organization, this exchange is the most efficient development, administration, transfer and implementation of knowledge assets [40]. Wilkesmann et al. [41] found that knowledge transfer has been studied using models, however, their use does not imply that knowledge can be found in its structure or that it is an exact replica by the initial receptor, in fact, knowledge is modified by the receiver. Some authors, i.e. [42] consider that knowledge transfer should be viewed as a reassembly process instead of a simple transfer-reception act where the key component is the medium the receptor uses to attain knowledge and then implement it [41].

In order for knowledge transfer to occur, knowledge needs to be identified as important, assimilated and commercially implemented [18]. Knowledge transfer can be viewed from different levels: inside and within organizational boundaries, among an organization's work teams or departments, or among organizations [43], furthermore, the professional training of a department's members influences knowledge transfer efficiency [44]. Data and information flow is the basis of knowledge transfer. Knowledge can be transfered in one-on-one interactions, or through information technologies [45]. Knowledge transfer is key during KM implementation, it is a necessity between employees and their co-workers [46]; success relies on individuals disposition to disseminate and encourage knowledge sharing [47].

KM should not be considered independent from business strategies [48], and must reflect the company's internal strategy that is valued by customers [49]. Knowledge can be found in repositories, however, large quantities can be made available tacitly; in an organization, transfer is defined as the process by which a unit (individual, team, department or section) is affected by the experience of another unit [50].

Organizations that effectively manage their knowledge will tackle new challenges more successfully [51]; if organizations do not know what knowledge they possess, the individuals that guard knowledge and processes cannot use them as a competitive advantage [52]. KM can be enabled through tools that do or do not make use of technology [53]. Additionally, KM support technologies must be coherent with a company's business processes [54].

Information, communication, and KM technologies are effective instruments that allow small and medium enterprises to remain afloat and grow in our current economy [55]; some of them collaborate with other enterprises as part of the supply chain and require a better integration of their data in terms of KM [56]. Since knowledge is substantially collaborative [3], several technologies have been used as knowledge support and management tools to facilitate the collaboration amongst individuals. In companies that provide professional services, survival relies on the exploitation of employee skills and knowledge [4], on knowledge's practical value for solving organizational issues [57], and on KM systems as mediums that enable knowledge sharing and development. Generally, technological tools imply the development of software projects that can be of high risk as the outcome may vary depending on their performance; information technology projects are a challenge for KM even with instructions on how to use them [58].

6.3 Software Strategy

The difficulty of implementing organization knowledge in an enterprise is a common scenario, particularly in small and medium enterprises (SMEs). It arises from: personnel changes or shift rotations, obsolete software and/or equipment, lack of communication mediums, and nonconsolidating knowledge and information; finding the ideal person for knowledge retrieval is both time and resource consuming and creates a dependence on this individual that is detrimental for the system. Figure 6.1 illustrates a company where departments require knowledge for planning and making decisions, and that might encounter limitations accessing the systems where it is stored.

When recovering stored knowledge (explicit), the technological systems used can be found to be obsolete, and even if they are compatible, it might be hard to identify in what server, equipment or document it is found.

Summarizing, knowledge is explicit and can have different presentations and be stored in various mediums; however, it might not be identified and taken advantage

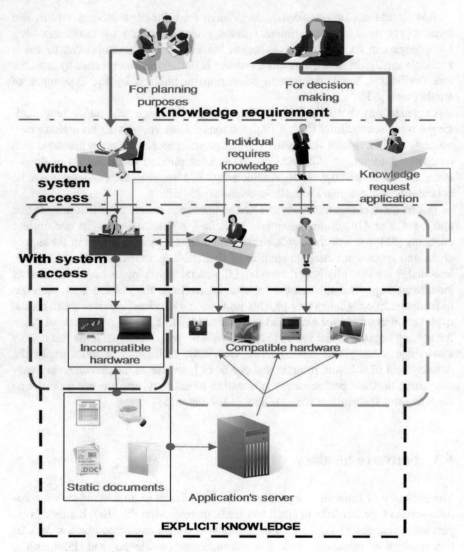

Fig. 6.1 Enterprise context

of. Information transfer needs to be improved; this would enable the exchange of relevant knowledge and improve a company's operation.

In order to solve these issues, and once the literature review was carried out, transfer enablers and theoretical framework tools were identified. The process requires several stages with specific tasks, at the end, an assessment is done to determine whether the objectives were met and the issues were solved.

Figure 6.2 shows the proposed model. It solves the knowledge transfer issues faced in the company and facilitates its replica. It consists of four stages:

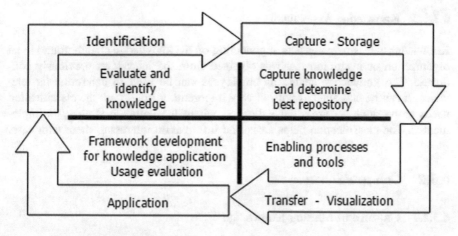

Fig. 6.2 Software strategy model

Identification, Capture/Storage, Transfer/Visualization and Application. Each stage is described in the next subsections.

6.3.1   Identification

This stage is the starting point for developing the system. It is the basis used for developing and increasing the size of the system. When knowledge is amply applied, there is an increase in work quality within an organization [59]. Knowledge is a base resource in enterprises and a key competitive advantage [60].

6.3.1.1   Knowledge Identification

It consists of the identification of the knowledge that will be shared. It determines the knowledge that needs to be transferred, this knowledge is already in the organization and adheres to the company's business strategy [61]. The following tasks need to be performed:

- Determine if knowledge is an asset in the main processes and adheres to the business strategy.
- Determine the media knowledge is stored in.
- Categorize the existent repositories.

The first two can be determined through research-oriented interviews of key personnel. For the third task, it is beneficial to include personnel working on information technologies.

6.3.1.2   Knowledge Assessment

Knowledge assessment consists of pondering on the knowledge already found in an organization or institution that was obtained from the interviews previously conducted. The knowledge shared by employees can be used to determine its relevance. It can be classified as critical, very important or important; the classification needs to be done by areas depending on where the software strategy is implemented. The classification helps areas find information interest to them with ease.

6.3.2   Capture/Storage

6.3.2.1   Capture of Missing Knowledge

Ideally, the knowledge that is used is already possessed and stored within an organization; however, the initial interviews might reveal knowledge has not been formally documented. In addition, there is a slim chance knowledge is stored in only one repository and that it is accessible, for these reasons the following tasks are proposed:

- Determine what repositories exist by analyzing the interviews of the IT personnel.
- Standardize the information and determine how it will be stored.
- Develop technological tools that facilitate accessing information, knowledge capture and knowledge storage.

6.3.2.2   Knowledge Repository

Knowledge needs to be stored in an accessible repository. It can be found as documents and archives; however, these are inefficient and take up a lot of space. In recent years, these archives have been digitized and are stored in accessible databases that can be transformed into repositories. In order for a repository to work correctly, its structure and requirements need to be in accord with an organization. Tools such as SQL can be used as the central engine of the database, and PHP or ASP as web platform generators. Preferably, the tools used are already found in the organization and satisfy its needs and requirements.

6.3.3   Transfer/Visualization

Once the information has been identified, assessed and sorted into repositories, the next stage is visualization. It refers to a simple representation of the knowledge and allows the use of the stored information.

6.3.3.1   Enablers

Knowledge can be disseminated through methodologies, procedures or processes that are involved in knowledge sharing within an organization [62]. In order for knowledge to be disseminated efficiently, it is important to take into account personal traits, skills, and motivations that influence its sharing and implementation. Knowledge transfer amongst employees and co-workers is key for a successful sharing initiative and other initiatives that are dependent on inter-employee knowledge sharing, thus making the organizational culture a key enabler.

6.3.3.2   Enabling Tools

The use of technological tools can facilitate knowledge sharing [53]. Once knowledge has been captured and stored in a repository, the intent is to implement a web based technological tool; it can be internet or intranet accessible and would provide an organization's users access to knowledge. This tool would facilitate time-consuming processes, enabling the development of new activities that might be beneficial for the organization. For this stage, it's encouraged to use a high level programming language, i.e. Java or C++, that ease module use and re-use. Beydoun et al. [63] suggest using ontologies for support during the development of information software because they provide domain knowledge and reusable software components for web applications.

6.3.4   Application

This stage refers to the launch of the technological platform, its advantages, the knowledge implementation by users with platform access, and knowledge's rendering and visualization. This stage is exemplified with shared document or repositories of "frequently asked questions" that facilitate dialogue and learning [52].

6.3.4.1   Usage Assessment

Finally, an evaluation is required on the software strategy used to share knowledge and on whether the knowledge transfer objectives were achieved. For the evaluation, the use of a survey on the initial interviewees is proposed. The survey would be descriptive of the methodology used in this research [47], and would capture the perceptions of the people involved; determining whether the objectives were met, i.e. whether the system has corrected the initial problem, and if not, what variants could ensure the proper functioning of the proposal.

Other elements to evaluate are: user adaptation to the new system in relation to their workload, the number of times information has been recovered from the system, the number of incidents that were raised in a timeframe and if they were responded to, information retrieval time for decision making support in administrative procedures, inter-departmental knowledge sharing, amongst others.

Once the technologies are implemented the evaluation will quantify the results. This step is necessary to maintain and improve the results [42], to determine if they are satisfactory or not, and finally to ensure the implementation of the software strategy represents an advantage or support for the organization.

6.4   Software Strategy Implementation

The software strategy was implemented in a pharmaceutical distribution company founded in 1992 that has grown in accordance to the market's needs over the years. Several procedures have remained unchanged since its foundation and now represent an operational problem, in particular for the commercialization department due to the technological advances and industry changes of the past 23 years.

The company markets biological medicines, alternative therapies and health preservation products. Company headquarters are located in northwestern Mexico, with branches in the center of the country, with approximately 80 individuals employed. The company must comply with Mexican tax and health risk regulations, thus, the interaction between the commercialization, purchasing, storage and research and development departments is required to ensure all norms and regulations are being followed.

Company departments perform specific tasks that require different knowledge, for this reason, some projects might require several departments to work together; e.g. when the commercialization department notices a drug needs restocking, it interacts with the storage and purchasing department to verify in-stock quantities, and collate a product's availability with customer preferences and arrival dates. This implies requests needing to be filled out, i.e. the information might not be readily available. However, if only these departments are communicating, and the vendor is unaware that some drugs are quarantined in the lab (therefore not communicating with this department), the quantities in stock might not be true, potentially affecting the credibility of the commercialization department and jeopardizing customer confidence.

Deficiency in spreading knowledge in a proper and timely manner between the marketing department and other departments triggers a series of situations that create problems of lost time, inaccurate or incomplete information, delayed shipments, etc. due to difficulties transmitting knowledge. Knowledge transfer is done on request when a department has a need for information, leading to lost time and re-work because the department with this knowledge might first need to be identified. Since its foundation, departments within the company interact directly between each other, and information has been stored in reports (e.g. laboratory reports), Excel spreadsheets or Word documents in delegated computers, or in the main commercial software that has been the base system for the commercialization department since 1996.

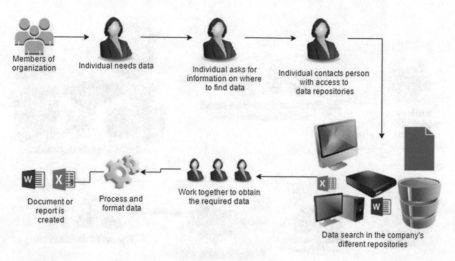

Fig. 6.3 Case study

There are information transfer deficiencies in the company that hinder knowledge transfer and its implementation. Explicit knowledge is either ignored or its existence is unknown. If a department requires it, as seen in Fig. 6.3, ignorance might lead to an inefficient information transfer among departments if the information that needs to be shared is not identified. Even though departments interact via emails, telephones or in person, there are still issues regarding relevant knowledge transmission. Most departments have stored data in plain text files using different software versions and hardware that might be damaged. The problem increases over time as staff retires and does not train replacements, leaving fewer capable people to capture and generate reports.

Figure 6.3 shows the process used in the company where the proposed software strategy was applied. First, an individual would need to identify the knowledge that they required. Then, they would identify where this information could be found and contact the personnel with access to the data repositories. Data would be found scattered and require its classification and evaluation. Both parties would work together to retrieve the desired information to then be able to process and format it, and finally generate an Excel or Word file that was stored in a single computer.

6.4.1 Identification

6.4.1.1 Knowledge Identification

It is the starting point to solve the company's problems. Figure 6.4 shows the framework used. The first step consisted on staff interviews to identify relevant knowledge, where it is found and in what repositories it is stored.

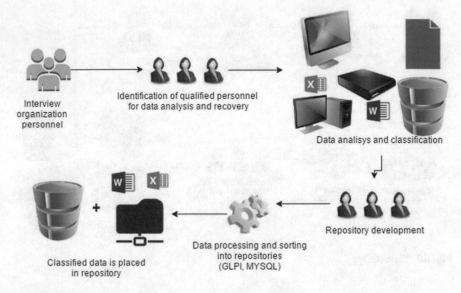

Fig. 6.4 Classification—availability

Key members of each department were interviewed; they were chosen based on their years of employment with the company, position and their interaction with the marketing department, which was the focal point of the study.

The interview consisted of a questionnaire with multiple choice and open questions, a description of the interview's objective and the areas where it would be applied. The questions were designed to obtain personnel's information regarding their: work experience, interactions with other departments, information handled, information required and the software tools used.

Figure 6.4 describes the proposed software strategy for this pharmaceutical distribution company. The first step consisted of interviewing personnel and identifying individuals qualified for data classification and recovery. This information is then analyzed and classified into categorized repositories where it can be stored.

6.4.1.2 Knowledge Assessment

The data analysis determined that most department knowledge is stored in individual digital documents. In the commercialization department, billing and telemarketing information is the most shared with telemarketing and management, and is related to customer preferences on shipment data, times or packages, and promotions; information on import permits are shared with the receiving, laboratory, telemarketing and purchasing departments. The analysis also found that the marketing department has a high interaction with all departments.

Interview information was captured in an Excel spreadsheet for a social network analysis and to learn how information and knowledge flow. NodeXL was used to analyze network graphs, capturing the relationship between employees and departments and providing an overview of their relationships. Knowledge flow was measured between departments, this helped identified those with less interactions that might be potential problem sources. In addition, NodeXL's visualization tool, makes departmental interactions easier to identify. The analysis yielded key information on major and minor interacting departments, and helped identify opportunities and weaknesses of knowledge transfer by determining the departments that required more attention.

6.4.2   Capture/Storage

In the second stage, knowledge that was previously identified needs to be stored, and missing knowledge needs to be captured and added to repositories as digital knowledge, making it more efficient and user friendly.

The interviews revealed the existence of a vast amount of explicit knowledge within departments and how it is stored in each area. The initial focus was on the reuse of explicit knowledge, however it was also necessary to know if users required extra knowledge, therefore, an additional question was added to the questionnaire. The general consensus was that departments did not require additional information; the processing area indicated the need for more instruction manuals that are found online, and marketing required information from the laboratory, human resources or administration on demand. The responses reaffirmed the need for explicit knowledge and its recovery. The solution was the development of an inter-departmental repository where information accumulated, knowledge was identified and new knowledge was developed. The repository could be physical, logical or a database depending on the type of information it would contain.

6.4.2.1   Knowledge Repository

Repositories were proposed according to the type of data they would store. The company made use of software acquired in 1992 that met the company's original requirements but had changed little over time. It had been adapted to meet current demands by installing third party software, e.g. for the generation of electronic invoices.

The language the aforementioned software was developed in is no longer in use and data was stored as encrypted plain text files, making damaged documents a challenge to recover with large data records stored in them over the years. The problem grew with the passage of time as the files increased in size and qualified personnel retired. When this study was performed, only a handful of people knew how to capture and generate reports. This software strategy seeks to recover old reports and import them to a database where knowledge can be accessed upon request. In addition to migrating to a database, using a common repository with identifiable .docx or .pdf files was also suggested, access control would be implemented as a security measure.

6.4.3   Transfer/Visualization

6.4.3.1   Enabling Processes and Tools

This stage refers to the repository where documents that are scattered throughout the company and were identified during the interviews will be consolidated. Open source software with access-controlled repositories exists, they are easy to install, require minimal configuration and can be instantly used; for these reasons, GLPI (Gestion Libre de Parc Informatique) software was chosen. This software also has tools that can be implemented posteriorly in the company, and supports in solving other problems.

Since process enablers refer to personnel's personal traits, skills and motivation on knowledge they possess, a final question was added to the interview. It sought to study staff's reaction and determine if there were obstacles for attaining what they desired.

Using a technological tool requires a computer with access to the company's internal network, which is why the interview also assessed if employees had access to computers. 100% of the interviewees have access to computers and the internal network, which ensures the use of the system in question. Figure 6.5 shows the enabling tool used for knowledge transfer, the system's interaction, its requirements and how explicit knowledge is obtained.

Interview results, interaction with the IT department and finding open source software tools enabled developing, implementing and assessing the software's usability.

Other open source software options include osTicket and GLPI. OsTicket is a ticket system for information technology companies with access control and allows databases, where documents and document descriptions can be added and categorized, and later be accessed by interested personnel. It is a basic and easy to use system.

GLPI is a more complete system with access control, incident control, an integrated help desk and where database cases and results can be presented as objectives, e.g. if a manual belongs to the accounting department, all accounting members can be granted access to it; it enables sharing with groups, individuals or with all members. It also has other useful business tools such as inventory control.

Based on the precious reasons, GLPI was selected. The installation file was downloaded from http://www.glpi-project.org/, and then unzipped in the main web

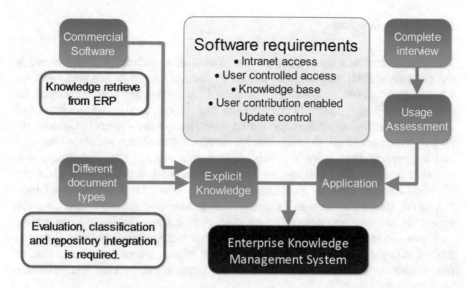

Fig. 6.5 Software interaction and requirements

server that was assigned by the IT department. Next, the directory was accessed from a web explorer with the following address http://dirlocalservidor/glpi/ thus beginning the automatic installation. The requirements were verified and no problems occurred during installation, thus the GLPI system was installed in the company's system with a general configuration and 4 users: glpi/glpi as administrator, tech/tech as technical support, normal/normal as a normal account and post-only/post-only for the post-only account.

Software installation was successful and the system was ready for use. However, information from the ERP commercial software needed to be recovered, task not performed by GLPI, for this reason an intermediate tool or middleware was developed. Its function was to recover ERP software content and insert it into a database from which data could be shared or added to documents that would later be added to the GLPI knowledge database.

The middleware was developed in Java. First, reports were generated in the commercial software, these data was then processed and validated by the middleware, and were temporarily allocated in the system until they are finally added to the database. The middleware was successfully implemented in shipment control software during the study.

MYSQL was the selected database as it was previously installed as GLPI software support. From the database, content could be shared or added to documents that would later be part of the GLPI knowledge database. Once application development and installation was completed the final step was personalizing and using the software.

6.4.4   Application

The selected tool was used in stages: first to recover knowledge that was stored in the company's ERP software, and second to manage, organize and make explicit knowledge available to authorized personnel. A framework was developed for using PHP software, specially created from this project, to access knowledge that had been identified, classified and stored; however, as the different stages of the project were performed, the framework changed. Open source software was used and an explicit knowledge recovery section was added. This knowledge was retrieved from the ERP software by Java based software that recovered plain text files and inserted them to a database where they could be easily accessed [64]. Figure 6.6 shows the knowledge application framework; an authorized user can access the software and retrieve information from the repositories and static files.

Figure 6.6 shows a framework suitable for new technological systems; it depicts the valorization and identification of common places where knowledge can be found. Knowledge then is stored in databases where it can be used and visualized by a company's users.

The application of the framework took place in stages. The intent was to recover knowledge that was frequently requested, i.e. there is an identified knowledge that was inaccessible. The proposed stages are:

The knowledge recovery application was developed using Java. The program retrieves text files generated in ERP software, processes them, and shows them in a reviewing window; once the information is corroborated, it is inserted into a

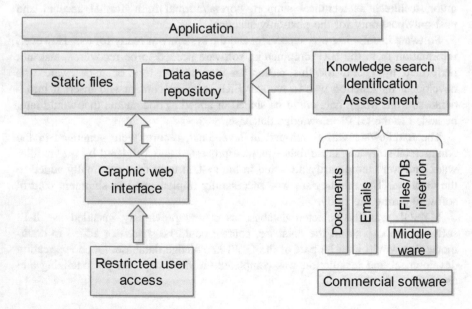

Fig. 6.6 Knowledge application framework

database where it can be consulted. This application was also used with a shipment program in the storage department and also integrated barcodes.

For static files, an evaluation was performed; it was decided that these files would be added to the database when were requested with the intent of avoiding filling the repository with outdated information. Instead, the IT department would keep lists with knowledge's location and only important knowledge would be added to the database from the start.

GLPI is open source web software in PHP that uses MYSQL databases; its features include: FAQs, databases where files can be stored and classified, enabled post-editing, content descriptions, and can restrict knowledge access to specific members. User control access is advantageous as it is simple and intuitive to learn.

GLPI and the Java software were used in collaboration to ensure the framework's requirements were met. GLPI provided access and control to the knowledge database and the Java middleware recovered already possessed knowledge. Between both they solve the company's problems and transfer knowledge efficiently without depending on third parties.

6.4.4.1 Usage Assessment

When assessing usage, GLPI includes metric tools for measuring usage; and, data related to the Java software can be retrieved by analysis the number of times knowledge has been recovered from the ERP software, this data represents the amount of information shared between the commercialization and warehouse departments.

Evaluating "Opened" and "Closed" events provides information on the number of incidents raised in a given period, how many were responded to and the time it took to answer them. Regarding the shipment software where the middleware was implemented, there were over 1000 shipments that required the used of the middleware in order to access commercialization and warehouse information. Personnel in the warehouse adapted to the software quickly as it reduced shipment revision times; they also stated that the software supported their work and allowed them to focus on other tasks that weren't previously done due to a lack of time.

Improvements on the shipping process are noticeable. Matrix based shipments are generally wholesale and have extensive bills that are manually reviewed twice and are very time consuming; another problem is that several drugs might present the same name but have different presentations, therefore, personnel need to review all contents with care.

This process was improved by introducing barcodes and by using the middle ware; which retrieves information from the commercial software, i.e. invoices and transfers, to then generate a report that can be inserted into the database. Task that can be done in a few clicks and that takes less than a minute. Once the information is loaded, a particular invoice can be selected and used to corroborate the contents of a shipment order using a barcode scanner. All scanned items are subtracted from the invoice list, and once all items are scanned, a report is generated.

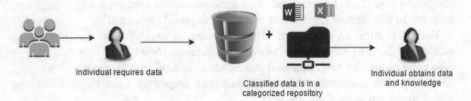

Individual requires data

Classified data is in a
categorized repository

Individual obtains data
and knowledge

Fig. 6.7 Knowledge usage sequence

An individual can perform loading and reviewing orders, while another person can be collecting invoice items to quicken the process. This enables a third person to focus on other activities and help by collect items, instead of manually corroborating the shipment contents for a second time; reducing the total time by 80%. Besides reducing shipment times, this system avoids shipment delays and presentation errors, and improves the department's efficiency.

From Fig. 6.7 we can see that once knowledge has been centralized and organized, retrieving specific information is easy for individuals as knowledge is organized in categorized repositories that are easy to access and identify. In other words, we went from the scenario in Fig. 6.4, to a much more reduced scenario in Fig. 6.7.

In order to corroborate that the system was in use within the company, a GLPI report was generated on the number of events and the time it took to resolve them. Figure 6.8 is a graph on the number of requests that were "Opened" and "Closed" between the months of April and July 2016; where most of the resolved requests were of knowledge requirements.

Figure 6.8 shows an increase in knowledge requests, which are a good indicator of the system's usage and user adaptability. The decline in knowledge requests is indicative of users no longer needing help to locate knowledge and being able to access it themselves. Originally, knowledge retrieval could take hours and involved regenerating multiple reports, hindering departmental efficiency. Implementing GLPI

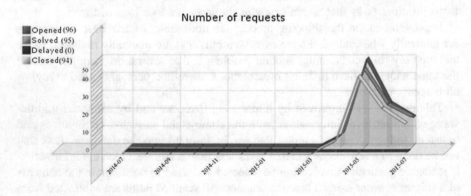

Fig. 6.8 Incidents report

allowed: reusing reports, identifying knowledge, reducing or eliminating department dependencies, reducing knowledge retrieval times and reducing lost time due to report reworking.

6.5 Conclusions

This research project was developed in a family company that granted access to its personnel for the interviews. The involvement of the IT department made it possible to implement and adapt all stages of the proposed software strategy. Interviews were made to key personnel who were willing to answer the questionnaire; their responses were analyzed and used to: identify what knowledge and information was available and where it was located, determine what each department considered key information, determine inter-departmental interactions, and to determine which departments were most important for knowledge transfer.

Implementing the software strategy solved the knowledge transfer problems that several departments faced in the company. Knowledge and static data was retrieved, classified and added to the GLPI software and the middleware provided support to the warehouse department by improving shipment activities and its functionality. Having knowledge centralized and organized, enabled personnel to locate knowledge in its repository and retrieve the desired information.

In addition, problems with explicit knowledge that had not been identified and was stored in other formats or software versions were solved. Nowadays, this information supports company performance, e.g. manual labor times was reduced for fulfilling shipment orders.

For future work or the replica of the proposed software strategy, retrieving knowledge in stages and determining possible gaps is suggested; e.g. [65] suggest using ontologies for analyzing system requirements and reducing communication errors or misinterpretations of customer needs. In this project, departmental knowledge that was passively retrieved was of great help as it identified informal or social interactions between company members. This can provide a lot of information, but precaution is needed, as it might not be useful.

It is feasible to replicate the model in other departments and to extend its application to other company branches, as it prevents knowledge from getting lost. According to [66] the incorporation and participation of knowledge from every stakeholder is essential in order to capitalize the collective intelligence.

This research contains a strategic model and a guide for its software implementation. GLPI software was used for the pharmaceutical distribution company, but similar software can also be used depending on the specific requirements of where it would be implemented. The strategy's and middleware's success depends on its continued use and growth; if it is no longer advantageous for the company it can be let go; therefore, its future depends on the company's personnel. For the time being, the system is yielding favorable results and continues to grow in the company.

References

1. Bagnoli, C., Vedovato, M.: The impact of knowledge management and strategy configuration coherence on SME performance. J. Manag. Governance **18**(2), 615–647 (2012)
2. Burmeister, A., Deller, J.: A practical perspective on repatriate knowledge transfer. J. Glob. Mobility. **4**(1), 68–87 (2016) . http://dx.doi.org/10.1108/JGM-09-2015-0041
3. Jones, P.M., Ph, D.: Collaborative knowledge management, social networks, and organizational learning. Syst. Soc. Internationalization Des. Aspects Hum. Comput. Interact. **2**, 306–309 (2001)
4. Criscuolo, P., Salter, A., Sheehan, T.: Making knowledge visible: using expert yellow pages to map capabilities in professional services firms. Res. Policy **36**(10), 1603–1619 (2007)
5. Wiig, K.M.: People Focus Knowledge Management. How effective Decision Making Leads to Corporate Success, 2004 edn. Elsevier (2004)
6. Alavi, M., Leidner, D.: Review knowledge management and knowledge management systems: conceptual foundations and research issues. MIS Q. **25**(1), 107–136 (2001)
7. Dalkir, K.: Knowledge Management in Theory and Practice. [en línea]. Taylor & Francis (2011)
8. Dehghani, R., Ramsin, R.: Methodologies for developing knowledge management systems: an evaluation framework. J. Knowl. Manag. **19**(4), 682–710 (2015). http://dx.doi.org/10.1108/JKM-10-2014-0438
9. Cricelli, L., Greco, M., Grimaldi, M.: Decision making in choosing information systems: an empirical study in Jordan. VINE **44**(2), 162–184 (2014)
10. Lee, C.H.S., Wong, K.Y.: Development and validation of knowledge management performance measurement constructs for small and medium enterprises. J. Knowl. Manag. **19**(4), 711–734 (2015). http://dx.doi.org/10.1108/JKM-10-2014-0398
11. McElroy, W.M.: The New Knowledge Management, pp. 1–23. Butterworth-Heinemann (2002)
12. Probst, G.J.B.: Practical Knowledge Management: A Model That Works. Prism-Cambridge Massachusetts, 17–30 (1998)
13. Reus, T.H., Lamont, B.T., Ellis, K.M.: A darker side of knowledge transfer following international acquisitions. Strateg. Manag. J. **37**, 932–944 (2016). doi:10.1002/smj.2373
14. Paramkusham, R.B., Gordon, J.: Inhibiting factors for knowledge transfer in info tech projects. J. Glob. Bus. Technol. **9**(2), 26–37 (2013)
15. Jeong, G.-Y., Chae, M.-S., Park, B.I.: Reverse knowledge transfer from subsidiaries to multinational companies: focusing on factors affecting market knowledge transfer. Can. J. Adm. Sci. **32**(2) (2016). doi:10.1002/cjas.1366
16. Sánchez de Gallardo, M., Nava Romero, M., Zulia, E.: Sistemas y barreras de la comunicación en institutos universitarios tecnológicos del municipio Cabimas, Estado Zulia, Venezuela. Enlace: revista Venezolana de Información, Tecnología y Conocimiento. **4**(3), 71–92 (2007)
17. Daghfous, A., Ahmad, N.: User development through proactive knowledge transfer. Ind. Manag. Data Syst. **115**(1), 158–181 (2015). http://dx.doi.org/10.1108/IMDS-07-2014-0202
18. Larkin, R., Burgess, J.: The paradox of employee retention for knowledge transfer. Employ. Relat. Rec. **13**(2), 32–44 (2013)
19. Churchman, W.C.: Managerial acceptance of scientific recommendations. Calif. Manag. Rev. **7**(1), 31–38 (1964)
20. Takeuchi, H., Shibata, T.: Japan Moving Toward a More Advanced knowledge Economy, 2nd edn. The world Bank, Washington, D.C. (2006)
21. Nonaka, I.: The knowledge-creating company. Harvard Bus. Rev. **69**(6), 96–104 (1991)
22. Hedlund, G.: A model of knowledge management and the N-form corporation. Strateg. Manag. J. **15**(S2), 73–90 (1994)
23. Nelson, R., Winter, S.: An evolutionary theory of economic change. Belknap Press, Cambridge, MA (1982)

24. Nonaka, I., Takeuchi, H.A.: The knowledge-creating company: how Japanese companies create the dynamics of innovation. Oxford University Press, New York (1995)
25. Polanyi, M.: The Tacit Dimension. Doubleday, Garden City, NY (1966)
26. Malhotra, Y.: Deciphering the Knowledge Management Hype. The Journal for Quality & Participation, July/August 1998 (Special issue on Knowledge Management). Published by the Association for Quality & Participation (1998)
27. Clark, P.A.: Organizations in Action. Competition Between Contexts. Routledge, London (2000)
28. Wiig, K.M.: Knowledge management: an introduction and perspective. J. Knowl. Manag. 1 (1), 6–14 (1997)
29. Debowski, S.: Learning and Development in a Knowledge Setting, in Knowledge Management. Wiley, Milton (2006)
30. Sveiby, K.E., Simons, R.: Collaborative climate and effectiveness of knowledge work—an empirical study. J. Knowl. Manag. 6(5), 420–433 (2002)
31. Davenport, T.H., Prusak, L.: Working Knowledge: How Organizations Manage What They Know. Harvard Business Press, Cambridge, MA (1998)
32. Liebowitz, J.: Key ingredients to the success of an organization's knowledge management strategy. Knowl. Process Manag. 1(1), 37–40 (1999)
33. Massey, A.P., Montoya-Weiss, M.M., O'Driscoll, T.M.: Knowledge management in pursuit of performance: insights from Nortel networks. MIS Q. 26(3), 269–289 (2002)
34. Nickols, F.: The tacit and explicit nature of knowledge, the knowledge in knowledge management. In: The Knowledge Management Yearbook, 2001, pp. 12–21 (2000)
35. Martín-de-, G.: BusinessReview exploring knowledge creation and transfer in the firm: context and leadership *. Universia Bus. Rev. 40, 126–137 (2013)
36. Jasimuddin, S.M., Connell, N., Klein, J.H.: Knowledge transfer frameworks: an extension incorporating knowledge repositories and knowledge administration. Inf. Syst J. 22(3), 195–209 (2012)
37. Yuan, Y., Lee, J.: Examining the role of knowledge transfer effect as a mediator variable among impact factors in knowledge innovation. Int. J. Bus. Inf. 7(2), 205–225 (2012)
38. Zapata Cantú, L., Rialp, J., Rialp Criado, C.A.: Generation and transfer of knowledge in IT-related SMEs. J. Knowl. Manag. 13(5), 243–256 (2009). http://dx.doi.org/10.1108/13673270910988088
39. Krishnaveni, R., Sujatha, R.: Communities of practice: an influencing factor for effective knowledge transfer in organizations. IUP J. Knowl. Manag. 10(1), 26–41 (2012)
40. Wiig, K.M.: Knowledge Management: an introduction and perspective. J. Knowl. Manag. 1 (1), 6–14 (1997)
41. Wilkesmann, U., Fischer, H., Wilkesmann, M.: Cultural characteristics of knowledge transfer. J. Knowl. Manag. 13(6), 464–477 (2009)
42. Szulanski, G.: The process of knowledge transfer: a diachronic analysis of stickiness. 82(1), 9–27 (2000)
43. Hansen, J.K.: How Knowledge is Transferred within the Danish Fashion Industry -Taking a Knowledge Management Perspective on the Creative Design Process. Master Thesis, Copenhagen Business School (2009)
44. Tasselli, S.: 2015. Social Networkd and Interprofessional Knowledge Transfer: The Case of Healthcare Professionals, Organization Studies 36(7), 841–872 (2015)
45. Sainio, H.: A dynamic Model for knowledge transfer and alliance learning in cross-border strategic alliances of software companies December (2007)
46. Mougin, J., Boujut, J.-F., Pourroy, F., Poussier, G.: Modelling knowledge transfer: a knowledge dynamics perspective. Concurrent Eng. 23(4), 308–319 (2015)
47. Rahman, R.A.: Knowledge sharing practices: a case study at Malaysia's Healthcare Research Institutes. Asia-Pacific Conference Library & Information Education & Practice, pp. 356–367 (2011)
48. Ritika, E.: Key enablers in the implementation of KM practices: an empirical study of software SMEs in North India 59–86 (2012)

49. Smith, A.D.: Knowledge management strategies: a multi-case study. J. Knowl. Manag. **8**(3), 6–16 (2004)
50. Argote, L., Ingram, P., Levine, J.M., et al.: Knowledge transfer in organizations: learning from the experience of others. **82**(1), 1–8 (2000)
51. Handzic, M., Lagumdzija, A., Celjo, A.: Auditing knowledge management practices: model and application. Knowl. Manag. Res. Pract. **6**(1), 90–99 (2008)
52. Mearns, M.: Knowing what knowledge to share: collaboration for community, research and wildlife. Expert Syst. Appl. **39**(10), 9892–9898 (2012)
53. Yuan, M.: An Integrated Knowledge Management Framework for Managing Sustainability Knowledge in The Australian Infrastructure Sector. Queensland University of Technology (2011)
54. Tsui, E.: The role of IT in KM: where are we now and where are we heading? J. Knowl. Manag. **9**(1), 3–6 (2005)
55. Majors, I., (2010). ICT and knowledge management models for promotion of SME's competitiveness. **6**(3)
56. Rodriguez-Enriquez, C.A., Alor-Hernandez, G., Mejia-Miranda, J., et al.: Supply chain knowledge management supported by a simple knowledge organization system **19**, 1–18 (2016). http://dx.doi.org/10.1016/j.elerap.2016.06.004
57. Zhang, W., Levenson, A., Crossley, C.: Move your research from the ivy tower to the board room. A primer on action research for academics, consultants, and business executives. Hum. Resour. Manag. **54**(1), 151–174 (2015)
58. Neves, S.M., da Silva, C.E.S., Salomon, V.A.P., et al.: Risk management in software projects through knowledge management techniques: cases in Brazilian incubated technology-based firms. Int. J. Project Manag. **32**(1), 125–138 (2013)
59. Nesheim, T., Gressgård, L.J.: Knowledge sharing in a complex organization: antecedents and safety effects. Saf. Sci. **62**, 28–36 (2014)
60. Sun, L.: Core competences, supply chain partners' knowledge-sharing, and innovation: an empirical study of the manufacturing industry in Taiwan. Int. J. Bus. Inf. **8**(2), 27 (2013)
61. Ruzo Sanmartín, E., Díez de Castro, J.A., Gómez Barreiro, M.: El ciclo evolutivo de la empresa familiar y las estrategias empresariales: estudio empírico de la empresa familiar gallega, pp. 484–497 (2003)
62. Cho, N., zheng Li, G., Su, C.: An empirical study on the effect of individual factors on knowledge sharing by knowledge type. J. Glob. Bus. Technol. **3**(2), 1–16 (2007)
63. Beydoun G., Low, G., Garcia-Sanchez, F., et al.: Identification of ontologies to support information systems development **46**, 45–60 (2014). http://dx.doi.org/10.1016/j.is.2014.05.002
64. Barcelo-Valenzuela, M., Carrillo Villafaña, P.S., Perez-Soltero, A., et al.: A framework to acquire explicit knowledge stored on different versions of software. Inf. Softw. Technol. **70**, 40–48 (2016)
65. Lopez-Lorca, A.A., Beydoun, G., Valencia-Garcia, R., et al.: Supporting agent oriented requirement analysis with ontologies **87**, 20–37 (2016). http://dx.doi.org/10.1016/j.ijhcs.2015.10.007
66. Alor-Hernandez, G., Sanchez-Ramirez, C., Cortes-Robles, G., et al.: BROSEMWEB: a brokerage service for e-procurement using semantic web technologies **65**, 828–840 (2014). http://dx.doi.org/10.1016/j.compind.2013.12.007

Chapter 7
GEODIM: A Semantic Model-Based System for 3D Recognition of Industrial Scenes

Yuliana Perez-Gallardo, Jose Luis López Cuadrado,
Ángel García Crespo and Cynthya García de Jesús

Abstract Keeping an inventory of the facilities within a factory implies high costs in terms of time, effort, and knowledge, since it demands the detailed, orderly, and valued description of the items within the plant. One way to accomplish this task within scanned industrial scenes is through the combination of an object recognition algorithm with semantic technology. This research therefore introduces GEODIM, a semantic model-based system for recognition of 3D scenes of indoor spaces in factories. The system relies on the two aforementioned technologies to describe industrial digital scenes with logical, physical, and semantic information. GEODIM extends the functionality of traditional object recognition algorithms by incorporating semantics in order to identify and characterize recognized geometric primitives along with rules for the composition of real objects. This research also describes a real case where GEODIM processes were applied and presents its qualitative evaluation.

Keywords Semantic rules · Recognized geometric primitives · Industrial scenes

7.1 Introduction

Describing in digital scenes what humans perceive by themselves is challenging, due to the large amount of information that must be handled in different contexts. Although several studies such as [1–10] have addressed 3D object recognition with

Y. Perez-Gallardo (✉) · J.L.L. Cuadrado · Á.G. Crespo · C.G. de Jesús
Computer Science Department, Universidad Carlos III de Madrid,
Av. Universidad 30, 28911 Leganés, Madrid, Spain
e-mail: yuliana.perez@alumnos.uc3m.es

J.L.L. Cuadrado
e-mail: jllopez@inf.uc3m.es

Á.G. Crespo
e-mail: acrespo@ia.uc3m.es

C.G. de Jesús
e-mail: cgarcia@alumnos.uc3m.es

© Springer International Publishing AG 2017
G. Alor-Hernández and R. Valencia-García (eds.), *Current Trends on Knowledge-Based Systems*, Intelligent Systems Reference Library 120,
DOI 10.1007/978-3-319-51905-0_7

excellent results, they may lack from semantic information to describe these objects.

According to [11], an inventory is a detailed, orderly, and valued relationship between elements that make up the assets of a company or person at a given time. It is detailed because the characteristics of each of the elements that integrate the patrimony are specified. It is considered orderly since it groups elements in their respective accounts. Finally, an inventory is valued because the value of every asset is expressed in units.

This research paper was carried out from an industrial approach. It focuses on the description of industrial scenes to create inventories of the elements that compose them. However, the analysis of such scenes may face certain difficulties, such as the size of the plant, the diversity of elements to recognize, or their amount. Also, in the real world, creating an inventory of the facilities of a factory implies several visits to the plant in order to identify and verify all the elements. Therefore, GEODIM aims to create a digital 3D mockup from a scanned 3D point cloud of a medium-large size industrial facility. This model includes all the information required to be familiar with all the types of objects involved and their features.

As proof of concept, GEODIM is applied to the 3D point cloud of the facility, and an enriched digital mockup with logical, physical, and semantic information is obtained as a result. The real case for this study involves the real indoor scene of an actual factory, and two evaluations are presented to assess the quality of classification of GEODIM.

The remainder of this research is thus structured as follows: Sect. 7.2 describes recent advances in the state of the art on object recognition and semantics in object recognition. Then, GEODIM is described in Sect. 7.3, while the real use case and the evaluation of the system are described in Sects. 7.4 and 7.5, respectively. Finally conclusions and future work are addressed in the Sect. 7.6.

7.2   State of the Art

Scenes of indoor factory facilities, building scenes, or even generation of product models, are examples of digital models in the industrial sector. Many efforts have been currently made to create these digital mockups in different contexts and with different objectives. Nevertheless, although their generation is a significant advance, they have failed to describe scenes from a semantic sense. The challenge now is therefore to describe the environment seen in a virtual scene in such a way that its physical and semantic properties are detailed.

Due to the nature of this research, the state of the art has been divided into a couple of subsections to describe the two main processes involved in GEODIM: object recognition and semantics in object recognition.

7.2.1 Object Recognition

Every method for 3D object recognition has special features depending on the domain or field where it is used. For instance, authors in [1] presented a technique for the recognition and reconstruction of surfaces from 3D data by applying line element geometry. Also, researchers in [2] modeled a gesturing hand through the use of key geometrical features of a hand and by constructing a skeletal hand model. Similarly, the work of [3] described a biometric-based security system using hand recognition. In general, the system relied on abductive learning and hand geometric features.

Other examples are the study presented by [4], where authors made use of color information to improve content-based retrieval of 3D models, and the work of [5], who incorporated highly discriminative affine invariant 3D information much earlier in the process of matching. Also, an approach for recognizing 3D objects was described by [6], where authors employed model synthesis to define a large number of possible geometric interpretations of images.

In addition, system identification was used for an emotion recognition model in [7]. The model included an extended Kohonen self-organizing map created by using a 26-dimensional facial geometric feature vector. On the other hand, researchers in [8] analyzed the detection effect of classic edge detection operators in infrared images. In the same year, a features recognition system was also proposed by [10], where the object-oriented structure was used for the generation of a geometric database. Finally, the work of [9] introduced a hybrid system that combined probabilistic graphical models with semantic knowledge for objects recognition in order to identify object in scenes for intelligent mobile robots.

Table 7.1 compares the different object recognition algorithms present in the literature. Although works [1, 8, 10] analyzed scenes and images by considering geometrical features of the objects and also detected existing objects, these studies failed to recognize such items. Similarly, in the area of Biometrics, despite positive

Table 7.1 Comparison of object recognition algorithms

Authors	Type of analysis			Semantics
	Reconstruction	Detection	Recognition	
[1]	✓		✓	✗
[2]			✓	✗
[3]			✓	✗
[4]			✓	✗
[5]			✓	✗
[6]			✓	✗
[7]			✓	✗
[8]		✓		✗
[9]			✓	✓
[10]		✓		✗

results obtained in the contexts of hand and emotion analysis, works [2, 3, 7] limited to object recognition without a semantic sense. They could have thus extended their functionality to meet extra information of the people identified based on semantic models. Meanwhile, works by [5, 6] considered the geometric features of objects as discriminant classification, while [4] relied on their color. However, none of these studies managed to provide a semantic meaning to those elements. Therefore, it is crucial to go beyond object recognition and expand the functionality of models, so they can describe a digital scene just as people perceive it. Similar to GEODIM, research by [9] relied on object recognition and used ontologies. However, the work made inferences by analyzing the context of the scenes supported by a knowledge base, such as: How long should a table measure so it can be considered as such? This type of knowledge is not applicable to GEODIM system, since tubes and objects from industrial scenes largely vary in size, although their shape is consistent. Therefore, identifying cylinders, tori, and spheres is valuable. GEODIM actually applies semantic rules to validate and correct the classification of objects and creates their topology. Furthermore, it is possible to extend its semantic meaning to connect to external ontologies.

7.2.2   Semantic on Object Recognition

As previously mentioned, GEODIM seeks to extend the object recognition process to enrich data obtained semantically in order to ease the inventory making process in factories. From this perspective, ten relevant works were found in the literature review within the scope of semantic technologies applied to 3D recognition.

First, authors in [12] proposed a feedback algorithm based on supervised feature extraction techniques; the algorithm used relevant feedback to retrieve semantically-similar objects. Also, a visual system was introduced by [13] to identify objects based on their functionality in an unknown terrain. Moreover, within the area of urban environments, a system for recognizing objects in 3D point clouds was described by [14]. The system recognized small objects in city scans.

A system for building object maps of indoor household environments was also developed by authors in [15], and the system relied on techniques for statistical analysis and feature extraction. Similarly, authors in [16] considered the spatial relationships between geometric primitives as part of the definition of the object and used ontological reasoning to solve the classification through SWRL rules. Likewise, a dataset called IAIR-CarPed was introduced by [17], which is the fine-grained and layered object recognition dataset with human annotations.

Recently, researchers in [18] developed a knowledge-based detection approach of 3D objects using the OWL ontology language known as WiDOP, which used VRML language to define the ontology of an indexed scene. In addition [19], introduced a framework that recognized materials on the surface of an object. The framework aimed to resolve the multi-label material recognition issue by exploiting object information. In that same year, a semantic mapping approach called

ASCCbot was also proposed by [20]. It relied on human activity recognition in a human–robot coexisting environment and enabled to create metric maps. Finally, researchers in [21] focused on the development of a framework that used top-down information to estimate the 3D indoor layout from a single image. The framework employed a semantic segmentation feature and an orientation map.

Table 7.2 compares algorithms of those studies that used semantics in object recognition. Although papers mentioned relied on semantics in different contexts, they limited themselves to labeling. On the other hand, GEODIM has the ability to provide semantic meaning to objects through the use of ontologies, so it is also possible to make inferences, obtain valuable information from the proposed model, or even go further, since the ontology of GEODIM can be linked to external sources of semantic information. On the one hand, Table 7.2 shows that, while studies from [12, 14, 15, 19] managed to identify elements within scanned scenes, they did not provide a semantic meaning to the segments. Consequently, they were unable to make inferences between elements or additional relevant information. On the other hand, not only does GEODIM make annotations to objects by means of tags, but it also provides these objects with a semantic meaning and describes their main features according to their type. From another perspective, research in [13, 20] created representation of spatial relations and maps by human-user and human–robot interactions, respectively. Likewise, GEODIM creates spatial relations of the objects by using Jena rules in order to define the topology of the elements of the scene. Moreover, its functionality is extended when the system adds extra information to objects in order to enrich the scanned scenes. The study by [21] was a much closer approach to our work, although the difference is that GEODIM analyzes scenes formed by clouds of points and does not train the model to classify. GEODIM also calculates the geometric characteristics of the objects, which allows it to migrate to other scenarios without needing so many changes in the core. The semantics is applied to infer topological position of the elements and correct classification problems. Moreover, the system provides the user with a better idea of the actual scene by having it fully available, instead of projecting a simple

Table 7.2 Comparison of semantic recognition algorithms

Authors	Type of analysis		Semantic by Tagging
	Recognition	Detect	
[12]	✓		✓
[13]	✗	✓	✓
[14]	✓		✓
[15]	✓		✓
[16]	✓		✓
[17]	✓		✓
[18]	✗	✓	✓
[19]	✓		✓
[20]	✗	✓	✓
[21]	✓		✓

incomplete image. With all this in mind, GEODIM innovates in terms of extending the operation of recognition algorithms reported in the literature by using the semantics of objects to improve their classification within a 3D point cloud.

This paper presents an extension of a conceptual model for the representation of digital mockups. The proposed model serves as a basis for the exchange of logical, physical, and semantic information of objects through a method of inverse engineering and by applying semantic technology and calculating spatial relationships. The model is able to recognize complex shapes from a 3D point cloud obtained from real objects in factories.

7.3 GEODIM Overview

GEODIM is an algorithm able to enrich models of indoor scenes of factories with logical, physical, and semantic information of the objects involved and by means of two processes: recognition of geometric primitives and semantic enrichment. These processes depicted in Fig. 7.1 are supported by a semantic model that contains all the information obtained through the execution of the algorithm. These processes and the model are described in the sections below.

Figure 7.2 depicts the workflow of GEODIM system, which starts by (1) receiving a point cloud of a real scene from a laser scanner. The process of geometric primitives recognition (:PrimitivesRecognition) analyzes the point cloud and then segments it in order to generate a simple classification of its elements (without semantic sense) and create a list of segments (2). From that point on, the semantic enrichment process calculates (3) specific properties according to their shape in (:GeometricFeatures), such as the trajectory, length, and diameter. Then, the list of (4) segments and their geometric features is sent to (:Topology) and (5) the spatial relationship of the segments are calculated by applying sematic rules. Afterward, the segments with their topology calculated by semantic rules are (6) sent to (:SemanticValidation). This task avoids issues of over- and under-segmentation of building primitives by (7) joining some segments or deleting others, so it is necessary to complete (8) the list of segments with spatial relationships and (9) recalculate in (:GeometricFeatures) the geometric features of the new segments formed. The result (10) is validated in (:ValidationByExperts), where expert users can (11) modify relationships as they deem appropriate. The list of modified-by-the-expert-user segments (12) is sent back to (:ValidationByExperts). Finally, all this calculated data are handled by (13) the Ontology Manager, which populates (14) the proposed ontology. The result would thus be a logical, physical, and semantic representation of a digital mockup obtained from real objects within an industrial environment. A more comprehensive description of the processes of the GEODIM algorithm is provided in the following section.

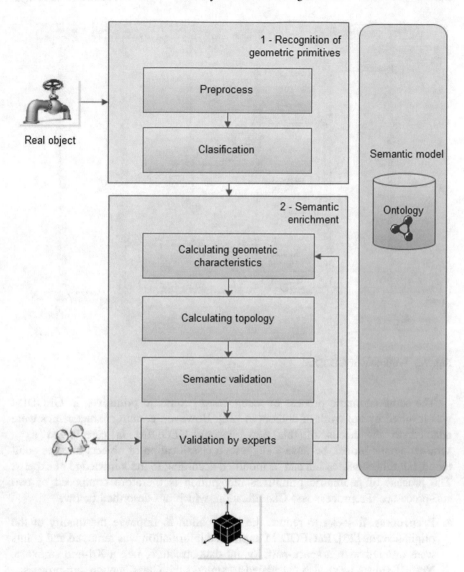

Fig. 7.1 Architecture of GEODIM

7.3.1 Process of Geometric Primitives Recognition

This process segments the point cloud according to its geometric characteristics. The entry of this process is the complete point cloud and the output is a list of objects segmented according to their geometric features. The geometric primitives analyzed by this process are cylinders, tori, and spheres.

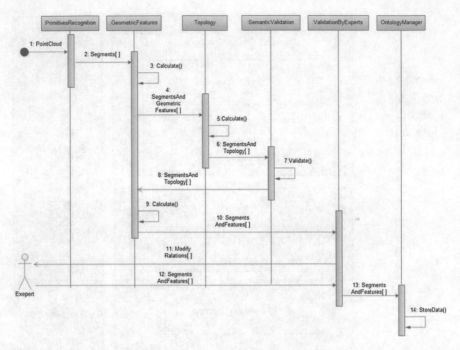

Fig. 7.2 Workflow of GEODIM

The semi-automatic process of recognizing geometric primitives in GEODIM was inspired by the work of authors in [22]. However, certain characteristics were adapted to the needs of this new system. GEODIM is categorized as a semi-automatic model, because it suggests a classification of objects into the point cloud, but this classification can be modified according to the knowledge of experts. The process of geometric primitives recognition is therefore composed of two sub-processes, Preprocess and Classification, which are described below.

a. **Preprocess**. It seeks to reduce the information to improve the quality of the original scene [23]. In GEODIM duplicated information was removed and points were ordered with a space-partitioning data structure, i.e., a Kd-tree structure. Normal values were also calculated to support the Classification sub-process.
b. **Classification**. It converts raw data into meaningful, useful, and understandable information [24]. Every object in GEODIM is classified with its corresponding geometric primitive according to its features, which are in turn obtained by calculating the percentage of fit of each primitive (cylinder, tori, sphere, and plane). These percentages are evaluated and the primitive type with the highest percentage of fit is selected and assigned to its corresponding object. The combination of geometric primitives allows for the generation of complex geometric shapes that describe real-world objects.

Since this research centers on industrial environments, every primitive has been assigned to a real element according to their similarity by semantic rules, which are described in the next section. Up to this point, GEODIM has a cloud of segmented points, and each segment is assigned to a class of primitive. In other words, GEODIM has created the logical representation of the scene. This list of segmented objects is the entry for the next process.

7.3.2 Semantic Enrichment Process

As entry, this process counts on a list of objects segmented according to their geometric characteristics, and as output, it provides with a logical, physical, and semantic representation of a digital mockup. To achieve this, GEODIM calculates the geometric properties of objects and their topologies, but it also semantically verifies the possible issues with over- and under-segmentation of building primitives by applying semantic rules.

Calculating geometric features and spatial relationships allows for the description of objects in a logical, physical, and semantic form. On the one hand, geometric features such as height, width, perimeter, and radius are considered logical information to GEODIM. On the other hand, spatial relationships and properties such as position, size, or number of points are viewed as physical information. Finally, semantic information for GEODIM involves matching every object with its corresponding element in the ontology proposed into the semantic model. However, merely the first two (logical and physical) tasks are calculated in the semantic enrichment process, while semantic information is described in "The topology of objects" [26].

a. **Calculating geometric characteristics**. This sub-process calculates specific geometric properties (trajectory, length, diameter, height, radius, etc.) according to the type of geometric primitive (cylinder, plane, torus, and sphere). All this information is stored based on the definition of the proposed ontology. Table 7.3 shows the geometric properties selected for every type of geometric primitive.

b. **Calculating topology**. Topology is a mathematical model used to define the location of and relationships between entities [25]. A topological 3D model should always be designed for specific requirements according to the application, due to its complexity and variation [26]. For this reason, GEODIM describes the

Table 7.3 Geometric properties allowed for every geometric primitive

Primitive	Number of points	Length	Height	Radius	Diameter	Minor radius	Major radius
Cylinder	✓	✓	✓	✓	✓		
Torus	✓	✓				✓	✓
Sphere	✓			✓	✓		
Plane	✓	✓	✓				

topology of objects by means of two concepts: spatial representation and spatial relationships via the design of semantic rules, depending on aspects such as the type of object and relationships permitted in the real world. To accomplish all assertions and constrains, different semantic rules were created and organized into three groups: (I) Spatial Relationship rules, (II) Spatial Representation rules, and (III) Redundant Shapes rules and Union of Shapes rules.

I. **Spatial representations**. In this case, "Meet, Overlap, Equal, Inside, and Contains" are the five spatial representations for two simple 3D objects without embedded holes selected for GEODIM. These are the most common relationships in industrial scenes. Table 7.4 shows the spatial representations that may exist in industrial scenes. For instance, to GEODIM spatial relationships between two pipes can only be of type Overlap, Equal, Inside, and Contains. However, the same pipe may have Overlap and Meet spatial representations with an elbow.

The assertions and constraints that describe the spatial relationships are verified in GEODIM by semantic rules, which were created and added to the semantic model. There is a rule for every combination according to the objects participating in the relationship. An example of this type of rules is described below.

A spatial representation Meet or Overlap of a pipe is valid if—and only if—the pipe has free connections, the Relatum is an elbow, a pipe, or a tee, and if there is a spatial relationship type Meet between them of at least 10% of their points.

SRP-Pipe-Elbow:

$$(?i \; rdf{:}type \; ont{:}SpatialRelation) \wedge (?i \; ont{:}has\_a\_relatum \; ?rel) \wedge$$

$$(?i \; ont{:}has\_a\_referent \; ?ref) \wedge (?rel \; rdf{:}type \; ont{:}Elbow) \wedge$$

$$(?ref \; rdf{:}type \; ont{:}Pipe) \wedge (?ref \; ont{:}number\_connections \; ?num) \wedge$$

$$lessThan(?num; \; 3) \wedge (?i \; ont{:}hassome \; ?adj) \wedge$$

$$(?adj \; rdf : type \; ont{:}Total\_Adjacency) \wedge (?adj \; ont{:} \; percentage \; ?per) \wedge$$

$$greaterThan(?per; \; 10) \wedge (?i \; ont{:}has\_some \; ?adjs) \wedge$$

$$(?adjs \; rdf{:}type \; ont{:}Side\_Adjacency) \wedge (?adjs \; ont{:}isValid \; ont : TRUE)$$

$$\rightarrow (?adj \; ont{:}isValid \; ont : TRUE)$$

Table 7.4 Spatial representations allowed for each entity

Object	Pipe	Elbow	Tee	Valve	Plane
Pipe	O, E, I, C	O, M	O, M		O, M
Elbow	O, M	E, I, C			
Tee	O, M		E, I, C	O, M	
Valve			O, M	E	
Plane	O, M				E

O Overlap, *E* Equal, *I* Inside, *C* Contains, *M* Meet

SRP-Pipe-Elbow rule describes the conditions for the relationship between a pipe and an elbow. The rule verifies conditions about spatial relationship, adjacency, and the adjacency type of this relationship. In this case, an elbow is necessary as Relatum and a pipe as reference. If the pipe has free connections, 10% or more should be the percentage of total adjacency in this relationship and a valid one.

II. **Spatial relationships**. Spatial relationships between objects are described according to the position of the first relative object called *Referent* towards intrinsic orientation of another object called *Relatum*. In a relative reference system, the relative position of a *Referent* toward its *Relatum* is described from the point of view of a third position called the *Origin* [13]. That is, the relative spatial position of an object depends on the viewing angle from which it is observed and its sensitivity to the rotation angle of the figure. Hence, for this model the *Origin* is equal to the position of the laser with respect to the scene. In GEODIM, the spatial relationships of objects are described by "Front," "Back," "Left," "Right," "Above," and "Below," a reference system of projective-relative relationships [27–30]. From this perspective, and considering that GEODIM is focused on the analysis of industrial scenes, elements to recognize have been limited to pipes, elbows, tees, and valves. Therefore, not all spatial relationships are permitted. Table 7.5 introduces the possible relationships for every *Referent* object according to its real features with its expected *Relatum*. For instance, a tube may have only two connections, which can be elbows or tees, while a valve has only one connection, since valves are usually only connected to tees.

Different spatial relationship rules were created, which are responsible for constructing spatial relationships between two objects called A and B. The following example shows the semantic rule for the relationship LEFT, considering that the remaining spatial relationships have similar behaviors.

Table 7.5 Spatial relationships for GEODIM	Referent	Possible relations	*Relatum* expected
	Pipe	2	Elbow, Tee, or Plane
	Elbow	2	Pipe
	Tee	3	2 Pipes and 1 valve, or 3 pipes
	Valve	1	Tee
	Plane	n	Pipe, or Plane

A connection type A Left B is valid if—and only if—the left points of Shape A are at least 90% close to the right points of Shape B.

SRL-Left:

$$(?i\ ont{:}percentage\ ?x) \wedge greaterThan(?x;\ 90) \wedge$$

$$(?i\ ont{:}refers\_to\_side\ ont{:}SIDE\_LEFT) \wedge (?i\ rdf{:}type\ ont{:}Side\_Adjacency)$$

$$\rightarrow (?i\ ont{:}isValid\ ont{:}TRUE)$$

SRL-Left rule describes the conditions for the relationship between a pipe and a tee. The rule verifies conditions about the adjacency of the relationship; 90% or more should be the percentage of adjacency in this relationship and an adjacency type = LEFT.

III. **Semantic validation**. Results from the recognition process of geometric primitives sometimes contain some over- and under-segmentation problems. GEODIM tries to avoid these issues through two types of semantic rules: Redundant Shapes and Union of Shapes rules. The former prevent over-segmentation by identifying the elements with the issue and joining them together as a single segment. The latter focuses on the same issue but under different conditions; it seeks the over-segmentation of objects that compose a real element. GEODIM has a Redundant Shapes rule for every type of combination of elements. The rule for two elbows is showed below as an example.

Combine segments A and B if—and only if—the Referent A and the Relatum B are of the same type, their centroids are at least 95% close, and all points are at least 85% close. That is, if A Equals B, A Inside B, or A Contains B.

RS-Elbow-Elbow:

$$(?i\ rdf : type\ ont : SpatialRelation) \wedge (?i\ ont : has\ a\ relatum\ ?rel) \wedge$$

$$(?i\ ont : has\ a\ referent\ ?ref) \wedge (?rel\ rdf : type\ ont : Elbow) \wedge$$

$$(?ref\ rdf : type\ ont : Elbow) \wedge (?i\ ont : has\ some\ ?adj) \wedge$$

$$(?adj\ rdf : type\ ont : TotalAdjacency) \wedge (?adj\ ont : percentage\ ?per) \wedge$$

$$greaterThan(?per;\ 85) \wedge (?rel\ ont : has\ some\ ?cd) \wedge$$

$$(?cd\ rdf : type\ ont{:}CentroidDistance) \wedge (?cd\ ont : percentage\ ?cdP\ erc) \wedge$$

$$greaterThan(?cdP\ erc;\ 95)$$

$$\rightarrow (?adj\ ont : isV\ alid\ ont : TRUE)$$

The RS-Elbow-Elbow rule describes the conditions for the relationship between two redundant elbows. The rule verifies conditions about spatial relationship, adjacency, and the centroids of the objects. Two Elbows as Relatum and as Reference are verified, 85% or more should be the percentage of total adjacency in this relationship, and the centroids between the two elbows should be 95% close.

Over-segmentation of segments belonging to the same cylinder has been observed as a result of primitives recognition algorithms. As previously stated,

GEODIM seeks to abolish these incorrectly classified segments and join the points in a single segment by means of Union Shapes rules:

Join segments if – and only if – the Referent A and the Relatum B are pipes and have a spatial representation Meet or Overlap.

US-Pipe-Pipe:

$(?i\ rdf : type\ ont : SpatialRelation) \wedge (?i\ ont : has\ a\ relatum\ ?rel) \wedge$

$(?i\ ont : has\ a\ referent\ ?ref) \wedge (?rel\ rdf : type\ ont : Pipe) \wedge$

$(?ref\ rdf : type\ ont : Pipe) \wedge (?i\ ont : hassome\ ?adj) \wedge$

$(?adj\ rdf : type\ ont : Total\ Adjacency) \wedge (?adj\ ont : percentage\ ?per) \wedge$

$greaterThan(?per;\ 10) \wedge (?i\ ont : has\ some\ ?adjs) \wedge$

$(?adjs\ rdf : type\ ont : Side\ Adjacency) \wedge (?adjs\ ont : isV\ alid\ ont : TRUE)$

$\rightarrow (?adj\ ont : isV\ alid\ ont : TRUE)$

The US-Pipe-Pipe rule shows the conditions for the relationship between two connected pipes. The rule verifies conditions of spatial relationship, adjacency, and adjacency side. Both the Relatum and the Referent are pipes, and 10% or more should be both the percentage of total adjacency in this relationship and a valid adjacency percentage.

c. **Validation by experts**. GEODIM is a semi-automatic model that suggests a classification of objects into the point cloud. In this sub-process an expert can modify items according to his/her knowledge. The expert can thus reclassify objects, delete them, or modify their properties. This sub-process was considered to support the quality of data classification.

7.3.3 Semantic Model

Semantic modeling helps define data in entities and their relationships. The set of entities of the proposed semantic model includes a taxonomy of classes used to support GEODIM in the representation of a digital scene as it is perceived in the real world. This approach is represented by an ontology. Figure 7.3 shows the semantic model proposed by GEODIM, which faithfully represents the behavior of the entities and their relationships in a real industrial scene, including assertions and limitations.

Table 7.6 describes the main elements of the proposed ontology.

The combination of geometric primitives allows for the generation of complex geometric figures that describe real-world objects. This ontology defines the objects (and their topology) recognized by the proposed model and it can link to external ontologies that will enrich the semantic meaning of such items.

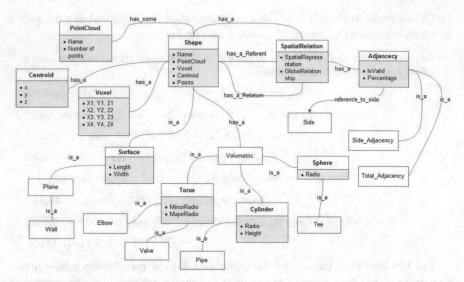

Fig. 7.3 GEODIM ontology

7.4 A Real Use Case: Objects Recognition in an Industrial Facility

This section describes the real use case that shows the functionality of GEODIM by explaining the stages of 3D reconstruction process. The problem is described below:

A great number of industrial facilities nowadays were built during the decades of 1960, 1970, or even 1980. Therefore, their plans have always been available in 2D, which makes them outdated and obsolete for today's needs due to continuous modifications, additions, or alterations that those establishments have suffered. As a result, the maintenance, reparation, and expansion of these places may become an extremely expensive, laborious, and—to a great extent—dangerous task. Since it is impossible to be completely familiar with the exact type, number, location, or dimension of objects, foremen and/or builders usually must leave the field and observe. This delays works and implies extra working hours. However, the solution to this issue is scanning inside the industrial facilities and recognizing existing objects with the help of GEODIM.

GEODIM was therefore employed in the use case for the creation of a digital 3D mockup from a scanned 3D point cloud of an industrial facility of medium-large size. Figure 7.4 represents the 360° view of a section of the industrial facility, which has 2,884,079 points. First, the shapes that GEODIM had to recognize for this studio were pipes and planes. Such information was useful to create more intelligent objects and enrich the 3D digital mockup.

The two main process of GEODIM, primitives recognition and semantic enrichment, are now described:

Table 7.6 Description of ontology

Entity	Description
PointCloud	The *PointCloud* object represents a set of points in a three-dimensional coordinate system, which describes a real-world scene
Centroid	It describes a point in the three-dimensional coordinate system that is located just at the center of an object
Voxel	It represents a grid in 3D space on the cloud of points. A Voxel object delimits, under spatial boundaries, a Shape belonging to the point cloud
Shape	It is a three-dimensional geometric object that occupies a place in space. A shape is classified as *Surface* or *Volumetric*. Every Shape element belongs to a particular *CloudPoint*
SpatialRelation	It describes spatial interactions between a *Referent* and a *Relatum*
Surface	It is a type of Shape describing an object that has length and width
Plane	It is a surface composed of points following a same direction. It is easy to find planar objects in the real world. Examples are tables, ceilings, walls, sidewalks, streets, and paper sheets
Volumetric	It describes a three-dimensional geometric Shape that has volume. A *Volumetric* element can be of type *Sphere*, *Torus*, *Cone*, or *Cylinder*
Sphere	It describes all elements having a spherical surface. Spheres can be found in both artificial environments and nature. A ball, a lemon, a doorknob, and a light bulb are examples of spheres
Torus	It is an element of the ontology describing a small circle that turns along a line drawn by another circle. Many real-life objects have a torus, such as donuts, a Roman cushion, a tire, or a light bulb
Cylinder	It describes elements formed by the simplest curves created by moving a straight line around the circumference of a circle. It is possible to find a cylinder in the real world in form of bottles, soda cans, columns, and pipes, for instance

Fig. 7.4 Industrial point cloud

7.4.1 Recognition Process of Geometric Primitives

In this process the scene was segmented according to its geometric properties by the primitive recognition algorithm (detailed description of this algorithm is beyond the scope of this paper). Table 7.7 summarizes segments obtained by the algorithm.

At this stage GEODIM has a list of classified elements belonging to the real industrial scene without semantic sense. That is, GEODIM has created the logical representation of the scene. The next step is the semantic enrichment process.

7.4.2 Semantic Enrichment Process

In this process GEODIM calculated extra information into describe the industrial scene with logical, physical, and semantic information by applying the following four sub-processes:

(a) **Calculating geometric characteristics**. GEODIM calculated specific geometric properties for every element of the list previously obtained. The system calculated the properties according to restrictions showed in Table 7.3, and it generated a list of elements classified with logical information.

(b) **Calculating topology**. The topology in GEODIM is defined by calculating spatial relationship, spatial representation, and a semantic validation of elements that belong to the point cloud by using semantic rules. The spatial relationship rules were applied for this use case to clarify the obtained result. An example of a created spatial relationships is presented: the relationships of plane_006 are detailed by a graphic and a scheme. In the real world, plane_006 represented the floor of the industrial walkway surrounded by two handrails; tubes represented the basis of these handrails. Figure 7.5 shows six spatial relationships created for plane_006 type ABOVE with 6 pipes: pipe_150, pipe_153, pipe_066, pipe_056, pipe_035, and pipe_033.

Next, the spatial representations (Meet, Overlap, Equal, Inside, and Contains) of elements were calculated by applying Spatial Representations rules. A visual example of the created spatial relationship is shown in Fig. 7.6, where a real industrial walkway is shown. This walkway was composed of different objects, six

Table 7.7 Classified segments

Size (number of points)	Primitive	Total segments	Segments classified correctly	Success rate (%)
2,884,079	Cylinder	112	103	86
	Planes	49	8	88
	Tori	0	0	0
	Total	161	111	68

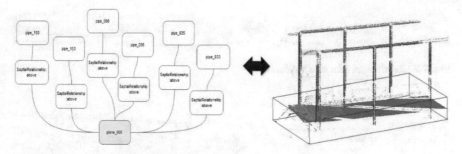

Fig. 7.5 Result of spatial relationships

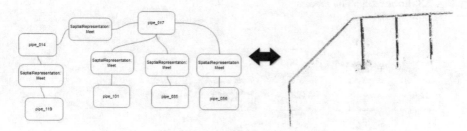

Fig. 7.6 Result of spatial representations

of which are depicted. Therefore, pipe_017 has four spatial representations type MEET with pipe_101, pipe_085, pipe_056, and with pipe_014, which in turn has a spatial representation with pipe_119.

Semantic validation tries to avoid issues in the process of geometric primitives recognition in terms of over- and under-segmentation. Figure 7.7 shows a sample of over-segmentation, where pipe_017 and pipe_075 stood for the same real object and had a spatial representation type EQUAL. Once the Semantic Validation rules were applied, the elements with a spatial representation type EQUAL were joined, creating only one element. In this example, the brand new element created was called pipe_u017.

Up to this moment, GEODIM always has an industrial scene with an extra definition of logical, physical, and semantic information. However, certain errors of classification or semantics may still exist. An expert validation is thus needed.

(a) **Validation by experts**. Expert users checked the elements in order to verify the classifications, properties, and the topology. They could correct any faults within the semantic model according to their experience. Figure 7.8 shows a classification error modified by the expert in the plane classification.

After all corrections, the semantic enrichment process was repeated until information was correct according to the viewpoint of experts. The final result of GEODIM was an industrial scene described with logical, physical, and semantic

Fig. 7.7 Result from semantic validation

Fig. 7.8 Error recognition process

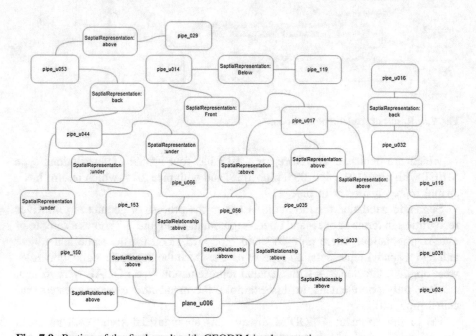

Fig. 7.9 Portion of the final result with GEODIM implementation

information. This model enabled the description of all objects into the industrial scene via semantic modeling. Figure 7.9 shows a portion of the final result after the use of GEODIM. The end result of GEODIM was the logical, physical, and semantic description of the elements that comprised the scanned point cloud. All this information allowed for the creation of the industrial inventory required by the factory.

7.5   Evaluation

Evaluating a geometric primitives recognition system is not an easy task. Literature carried out for this research managed to describe and carry out different types of evaluations, although these cannot always be useful for other systems. Algorithms evaluation actually varies depending on diverse factors, such as the amount of points in the point cloud, density, or the type of elements involved. However, from a general perspective, two different methods are possible to assess software and tools:

(a) Quantitative evaluation methods. They are based on the assumption that the software product has at least a measurable property that can change as a result of using the methods/tools to be evaluated. Quantitative evaluations can be developed in three different ways: case studies, formal experiments, and surveys. For instance, authors in [31] introduced experiments to validate the effectiveness of their proposed QDFT descriptor under geometric transformation. Similarly, the one-versus-rest multiclass classification experiments was addressed by [32], where true and false samples were labeled differently. Authors used accuracy rates to evaluate the performance of different methods. Furthermore [33], proposed a performance comparison of their work and presented seven other studies. The comparison aimed at measuring average precision. Finally, authors in [9] discussed the success of object recognition of the research, and different aspects were measured, such as the inclusion of contextual information about the objects, their geometric and appearance features, and their classification based on their type. Metrics used were precision and recall.

(b) Qualitative methods. The term Feature Analysis is used in literature to describe a qualitative evaluation, which is based on (1) identifying the requirements that users have for a particular task or activity and (2) mapping those requirements to features that a method/tool aimed at supporting that task/activity should possess. An example of this qualitative assessment is the qualitative comparison provided by [34], where two methods were analyzed. In total, 31 clusters were extracted out of the computed hierarchy for a particular model. Then, results from the two methods were compared to determine which one managed to generate the best stylized model.

In the end, a quantitative approach was selected to evaluate the quality of classification of GEODIM. Measurements employed were: Segments Classified and Segments Classified Correctly; i.e., true positive (TP), false positive (FP), false negative (FN), precision (P), and recall (R). Two evaluations were proposed. The former was used for the classification results in the process of geometric primitives recognition, whilst the latter was carried out for the final results of GEODIM in order to compare results and demonstrate how GEODIM improved the traditional classification in an industrial point cloud.

Segments Classified referred to the number of segmented elements belonging to the original point cloud, while Segments Classified Correctly or TP concerned those segments correctly classified according to their geometric properties. Similarly, FP stood for segments incorrectly classified, whilst FN comprised segments not classified based on their corresponding geometric primitive. Also, in this case P and R were defined as a set of classified segments and a set of relevant segments. That is, P referred to the fraction of classified segments relevant to the process, while R indicated the fraction of the segments relevant to the process but which were also successfully classified.

$$P = TP/(TP + FP). \tag{7.1}$$

$$R = TP/(TP + FN) \tag{7.2}$$

Table 7.8 shows the result from the process of geometric primitives recognition. The point cloud was segmented in 161 elements, most of which were cylinders and planes, in addition to some tori elements. The table shows low precision and recall metrics, since the algorithm misclassified 32 elements as tori. In the case of cylinders, precision P was 0.86, whilst planes had a precision of 0.88, although its recall measure was low due to 41 elements that were incorrectly classified. In total, this method had a precision of 0.69 and a recall of 0.68.

The implementation of the semantic enrichment process of GEODIM improved these results, such as Table 7.9 shows. It can be observed from the new table that precision and recall for cylinders and planes became 100%. However, certain

Table 7.8 Results from object recognition process

S	Primitive	TS	TP	FN	FP	P	R
2,884,079	Cylinders	112	103	9	16	0.86	0.92
	Planes	49	8	41	1	0.88	0.16
	Tori	0	0	0	32	0	0
Total		161	111	50	49	0.69	0.68

S Size of point cloud, *TS* Total of segments, *TP* True Positive, *FP* False Positive, *P* Precision, *R* Recall

Table 7.9 Results from semantic enrichment process

S	Primitive	TS	TP	FN	FP	P	R
2,884,079	Cylinders	84	84	0	0	1	1
	Planes	41	41	0	0	1	1
	Tori	0	0	0	18	0	0
Total		125	125	0	18	0.87	1

S Size of point cloud, *TS* Total of segments, *TP* True Positive, *FP* False Positive, *P* Precision, *R* Recall elements were still misclassified, since they were sets of points that did not fit any geometric primitive analyzed. This can be considered as lost information, and, for this reason, precision in the process was 87%. The table also shows that the total number of segments decreased due to the semantic validation process, wherein the over-segmentation was solved.

7.6 Conclusions and Future Work

This research introduced GEODIM, a semantic model-based system for the recognition of industrial scenes that creates a conceptual model for semantic representation of digital mockups (i.e., a semi-automatic inventory). GEODIM enables users to enrich models of indoor scenes of factories with logical, physical, and semantic information by using a semantic model and applying two processes: geometric primitives recognition and semantic enrichment.

Geometric primitives recognition actually classifies the elements of the scanned scene according to their geometric characteristics, while the semantic enrichment process calculates valuable information for every item, including their topology, by using semantic rules. These rules describe the assertions and restrictions of behavior in the real world of industrial elements used (pipes, planes, elbows, and valves). To validate the functionality of GEODIM, the real use case of an industrial facility was thus introduced, and the research also addressed two quantitative evaluations in order to show the classification quality of GEODIM.

Results obtained showed that GEODIM does improve the classification of objects with 87% of success in this case, although elements incorrectly classified remained present due to under-segmentation issues. Future work will thus seek to include the composition of complex industrial elements based on the combination of several recognized objects in order to improve the recognition of a wide range of both simple and complex real industrial objects. Further research will also aim at including automatic element recognition to improve the quality of the classification and enrich industrial scenes. Mechanisms to avoid under-classification of elements will also be pursued.

Acknowledgements This work was supported by the National Council of Science and Technology of Mexico (CONACYT) and the Public Education Secretary (SEP).

References

1. Hofer, M., Odehnal, B., Pottmann, H., Steiner, T., Wallner, J.: 3D shape recognition and reconstruction based on line element geometry. Proc. IEEE Int. Conf. Comput. Vis. **II**, 1532–1538 (2005)
2. Bhuyan, M., Neog, D., Kar, M.: Hand pose recognition using geometric features. In: Communications (NCC), 2011, pp. 0–4 (2011)

3. El-Sayed, M., Radwan, E., Zubair, A.: Abductive neural network modeling for hand recognition using geometric features. In: Neural Information Processing, pp. 593–602 (2012)
4. Pasqualotto, G., Zanuttigh, P., Cortelazzo, G.M.: Combining color and shape descriptors for 3D model retrieval. Signal Process. Image Commun. **28**(6), 608–623 (2013)
5. Soysal, M., Alatan, A.A.: Joint utilization of local appearance and geometric invariants for 3D object recognition. Multimed. Tools Appl. **74**(8), 2611–2637 (2013)
6. Hejrati, M., Ramanan, D.: Analysis by synthesis: 3D object recognition by object reconstruction. In: 2014 IEEE Conference on Computer Vision and Pattern Recognition, pp. 2449–2456 (2014)
7. Majumder, A., Behera, L., Subramanian, V.K.: Emotion recognition from geometric facial features using self-organizing map. Pattern Recogn. **47**(3), 1282–1293 (2014)
8. Junyan, L., Qingju, T., Yang, W., Yumei, L., Zhiping, Z.: Defects' geometric feature recognition based on infrared image edge detection. Infrared Phys. Technol. **67**, 387–390 (2014)
9. Ruiz-Sarmiento, J.-R., Galindo, C., Gonzalez-Jimenez, J.: Scene object recognition for mobile robots through Semantic Knowledge and Probabilistic Graphical Models. Expert Syst. Appl. **42**(22), 8805–8816 (2015)
10. Nasr, E.S.A., Khan, A.A., Alahmari, A.M., Hussein, H.M.A.: A feature recognition system using geometric reasoning. Procedia CIRP **18**, 238–243 (2014)
11. Gaither, N., Frazier, G.: Administración de producción y operaciones (2000)
12. Leifman, G., Meir, R., Tal, A.: Semantic-oriented 3d shape retrieval using relevance feedback. Vis. Comput. (2005)
13. Hois, J., Wünstel, M., Bateman, J., Röfer, T.: Dialog-based 3D-image recognition using a domain ontology. In: Spatial Cognition V Reasoning, Action, Interaction (2007)
14. Golovinskiy, A., Kim, V.G., Funkhouser, T.: Shape-based recognition of 3D point clouds in urban environments. In: 2009 IEEE 12th International Conference on Computer Vision, no. ICCV, pp. 2154–2161 (2009)
15. Rusu, R., Blodow, N.: Close-range scene segmentation and reconstruction of 3D point cloud maps for mobile manipulation in domestic environments. In: Intelligent Robots and Systems (2009)
16. Günther, M., Wiemann, T.: Model-based object recognition from 3d laser data. In: KI 2011 Advances in Artificial Intelligence (2011)
17. Wu, Y., Liu, Y., Yuan, Z., Zheng, N.: IAIR-CarPed: a psychophysically annotated dataset with fine-grained and layered semantic labels for object recognition. Pattern Recognit. Lett. **33** (2), 218–226 (2012)
18. Hmida, H., Cruz, C., Boochs, F., Nicolle, C.: Knowledge base approach for 3d objects detection in point clouds using 3d processing and specialists knowledge. arXiv Prepr. arXiv1301.4991 (2013)
19. Yang, L., Xie X.: Exploiting object semantic cues for Multi-label Material Recognition. Neurocomputing **173**, 1646–1654 (2015)
20. Sheng, W., Du, J., Cheng, Q., Li, G., Zhu, C., Liu, M., Xu, G.: Robot semantic mapping through human activity recognition: a wearable sensing and computing approach. Robot. Auton. Syst. **68**, 47–58 (2015)
21. Park, S.-J., Hong, K.-S.: Recovering an indoor 3D layout with top-down semantic segmentation from a single image. Pattern Recognit. Lett. **68**, 70–75 (2015)
22. Attene, M., Patane, G.: Hierarchical structure recovery of point-sampled surfaces. Comput. Graph. Forum **29**(6), 1905–1920 (2010)
23. Gonzalez, R.C., Woods, R.E.: Digital Image Processing, 3rd edn. (2006)
24. Mountrakis, G., Im, J., Ogole, C.: Support vector machines in remote sensing: a review. ISPRS J. Photogramm. Remote Sens. **66**, 247–259 (2011)
25. Liu, L., Zsu, M.: Encyclopedia of Database Systems (2009)
26. Zlatanova, S., Rahman, A.A., Shi, W.: Topological models and frameworks for 3D spatial objects. Comput. Geosci. **30**(4), 419–428 (2004)

27. Moratz, R., Nebel, B., Freksa, C.: Qualitative spatial reasoning about relative position. In: Spatial Cognition III (2003)
28. Moratz, R., Tenbrink, T., Bateman, J., Fischer, K.: Spatial knowledge representation for human-robot interaction. In: Spatial Cognition III (2003)
29. Méndez, V., Rosell-Polo, J., Sanz, R.: Deciduous tree reconstruction algorithm based on cylinder fitting from mobile terrestrial laser scanned point clouds. Biosyst. Eng. **124**, 78–88 (2014)
30. Levinson, S.: Frames of reference and Molyneux's question: crosslinguistic evidence. Lang. Space (1996)
31. Li, H., Liu, Z., Huang, Y., Shi, Y.: Quaternion generic Fourier descriptor for color object recognition. Pattern Recognit. **48**(12), 3895–3903 (2015)
32. Hong, C., Yu, J., You, J., Chen, X., Tao, D.: Multi-view ensemble manifold regularization for 3D object recognition. Inf. Sci. **320**, 395–405 (2015)
33. Rubio, J.C., Eigenstetter, A., Ommer, B.: Generative regularization with latent topics for discriminative object recognition. Pattern Recognit. **48**(12), 3871–3880 (2015)
34. Attene, M., Falcidieno, B., Spagnuolo, M.: Hierarchical mesh segmentation based on fitting primitives. Vis. Comput. **22**, 181–193 (2006)

Chapter 8
Beyond Interoperability in the Systems

Aleksander Lodwich and Jose María Alvarez-Rodríguez

Abstract Future complex products will be different from existing ones in several relevant ways. They will be more intelligent and connected and they will have to be greatly leaner across software and hardware in order to handle safety, security and resource demand. *Industrial Internet*, *Industry 4.0* and *Internet of Things* will greatly shift responsibility for products away from engineering departments towards the actual environment in which the products are employed. This situation will eventually transform the most tested companies into intelligent building platforms where the responsibility for designing, producing and delivering will be distributed among market parties in unrecognizable ways. The benefits of these upcoming changes will be higher utility of products for customers and new levels of production flexibility and efficiency. However, this new environment can only be attained if developmental and operational platforms and embedded products can rely on reusing the explicit knowledge used in their designs. The provision of technology for this new environment goes far beyond asking for *tools interoperability*. In this chapter a conceptual layer of interoperability is outlined describing what kind of features a powerful new interoperability technology should support in order to fuel desired changes in engineering and production paradigms.

Keywords Internet of tools · Interoperability layer · Internet of things · Industrial internet · Technology · Industry40

8.1 Introduction

Technology has been always considered a driver of change. More specifically, digital technology (hereafter the term technology refers to digital technology) fueled by software is currently embedded in any task, activity, process that is done in any organization or even in our daily life activities. This new digital age also

A. Lodwich · J. María Alvarez-Rodríguez (✉)
Department of Computer Science, Carlos III University of Madrid, Madrid, Spain
e-mail: joalvare@inf.uc3m.es

© Springer International Publishing AG 2017   161
G. Alor-Hernández and R. Valencia-García (eds.), *Current Trends on Knowledge-Based Systems*, Intelligent Systems Reference Library 120,
DOI 10.1007/978-3-319-51905-0_8

implies that industries and businesses need to reshape their strategies to become part of the new "Industrial Revolution", the Industry 4.0 [1, 12, 27].

They need to understand this challenging environment in which customer needs and behaviors are continuously changing. The use of emerging technologies must then serve to fulfill the new needs but also to create new technology-driven business opportunities. However, organizations must not focus only on technology but people: consumers, workers and partners. The first step to become a leading digital company will rely on changing the corporate culture to incorporate technology as a key driver to empower people and to drive the change and disruption in a particular sector.

In this context, the "2016 Accenture Technology vision report" [22] outlines five emerging technology trends that aim to shape this new environment. More specifically, organizations will create a new wave of data-driven solutions powered by artificial intelligence ("Intelligent Automation"). It will be completely necessary to equip people with the required skills to select the proper technology ("Liquid Workforce") and to find the proper (e.g. including quality factors such as scalability, usability, security, etc.) execution platforms ("Platform Economy") boosting disruption ("Predictable Disruption"). However, this new environment will also raise new risks that must be addressed to create a real and trustworthy digital ecosystem ("Digital Trust").

In the field of engineering, the Industry 4.0, "Industrial Internet", "Industrial Data Spaces" and many other initiatives are trying to establish guidelines, methodologies, good practices to drive this new concept in a proper way. In general, the application of digital technology to the core of a sector will generate innovation improving the way of working and easing the transition to a digital environment. To do so, a set of well-grounded technologies (inspired by the "Technology Radar", Gartner Hype Cycles and many other consultancy reports) will be necessary and may include: Security, Big Data, Mobility, Natural Language Processing, Internet of X (Things, People, Tools, Everything, etc.), User Interfaces, 3D Printing, Virtual Reality or Cloud Computing to name just but a few.

In this context, last times have seen the creation of a new concept, Cyber-physical systems [19] (CPS), to name those engineering systems aimed to combine technologies coming from different knowledge areas such as mechanical, electrical and software engineering. Cyber-physical systems are considered the next type of systems in the field of safety critical systems. Cars, planes, railways, medical devices, robotics systems are examples of the combination of different engineering areas governed by software to provide interconnected systems that can communicate and collaborate each other to provide different behaviors and interactions with humans and machines. The 4th Industrial Revolution and the resulting industrial era imply unprecedented levels of interaction and coalescence between models and formalisms and this not only in different domains of systems engineering but also between systems organizing production, distribution, support and decommissioning.

However, in the era of the 4th Industrial Revolution, the development of CPS has still to face existing challenges in the development of complex systems.

Although digital technology represents a big step towards the creation of collaborative development environments, engineers must be equipped with the required skills to become part of the digitization of the development processes. From the inception of the system to the production environment, everything is connected, e.g. a requirement is implemented by a model, a physical model that governs a function of the system, is validated by a test case or an evidence is used to certify a CPS by a certification authority, etc.

Tomorrow's engineering is not anymore a set of vertical activities developed by different engineers but a collaborative process in which people and technology is completely involved in the engineering process. It is clear that technology apart from being a key enabler for this new environment, can also imply the need of collaboration between tools to enable engineers to produce timely products and services in terms of cost, time and effort. This situation also means that it is necessary to provide the proper mechanisms to reuse the knowledge embedded [15] in the different system artifacts and to share them along the different stages of the development life cycle. Currently, Product Line Engineering (PLE) and Software Product Lines [23] are gaining attraction to develop such complex systems easing the configuration activities through the management of commonalities and variabilities. On the other hand, interoperability is again a major challenge to be addressed at different levels since technology stacks also require more powerful communication and interpretation mechanisms.

In general, it has been found that such radical rethink, as it is stated for the 4th Industrial Revolution, is not only good to save costs in the production of complex products. It seems that an efficient engineering process regarding system products and the necessary production facilities can be the cornerstone to reconcile demand for attractive products and resources. Thus, the contemporary global economy needs to reshape its strategy to avoid excessive use of natural resource and to provide sustainable industries. Sustainability is regarded as high priority by more and more industrial nations for which ecological footprint is becoming a real pressing issue.

Sustainable "green" industries must produce less waste, create longer lasting products, reuse more (not only material but also whole systems), reduce involved transportation in order to achieve this major objective of providing "greener" engineering environments. This can only be achieved by producing more intelligent, versatile, continuously upgradeable niche-optimized components which are produced and disposed off in the neighborhood—an industrial strength steer away off industrialization, so to say. At the end of this process, industry will be completely immersed into an augmented natural eco-system and, at the same time, it will be possible to acquire some relevant properties like high recycling rates, high levels of customization and object intelligence as it is known from living things.

In this work, authors review the need of interoperability in the Industry 4.0 and outline a conceptual interoperability layer for the industry sector as a mean to produce and operate Cyber-physical systems in a safer, greener and timely manner.

8.2 The Challenge of Interoperability in the Industry 4.0

This new challenging environment, affecting businesses, governments, people and economy implies a technological upgrade to be accomplished in accordance with the six design principles [6]: interoperability, virtualization, decentralization, real-time capability, service orientation and modularity.

In context of interoperability, it is necessary to ensure seamless communication between development tools ("Internet of Tools") and among resulting machines ("Internet of Things") Practical experience shows that there is no distinct boundary between the two and that technology should support a continuum of man-toolmachine interactions.

Since the amount of engineering time necessary to generate individualized products is a key cost driver, acceleration of engineering by employing sophisticated knowledge composition strategies seems to be the most promising path to cost reduction aside of achieving shorter time to market. Although some engineering companies are dedicated to react to this need, the reality shows that existing engineering and knowledge generation/consumption processes must be reorganized and technologically bolstered to deliver the required speed—none of which is clear how to do. As an example of the ongoing research, in the CRYSTAL[1] project, a good number of technology providers and final customers/consumers have been involved into making the best of current engineering environments by improving interoperability between system engineering tools. After three years, a complete exercise of rethinking the existing engineering processes has been made to support the future of industrial developments.

From this new mindset, it has been found that a new set of technologies and ways to use them can revolutionize the efficiency of engineering and production and can enable completely new ways of interacting with customers/consumers. In regards to OEMs (Original Equipment Manufacturer), suppliers, producers or consumers, they could be completely replaced with new terminology because the current terms reflect a certain path of endowing objects with knowledge ("*designing products*").

From a more sophisticated vision, this design endowment will be much stronger distributed among involved parties and the idea of an OEM could be much better suited to the customer/consumer than to some company. It can be seen as the *democratization of industry*, a kind of a new sharing economy, the *sharing Industry 4.0*. This vision aims to not only good save costs but to deliver an efficient engineering process regarding system products and the necessary production facilities being be a cornerstone to reconcile demand for attractive products and resources.

In order to address such goals, it inevitably leads to restoring decentralized production goods, distribution, maintenance and disposal of systems which would involve lifecycle management from cradle to cradle for almost any kind of object

---

[1]Critical Systems Engineering Acceleration Project of ARTEMIS JU, http://www.crystalartemis.eu/.

produced between market participants who did never even guess they would have to interact or rely on each other a priori—and given a global economy, this will occur at a global scale. What we claim by this is that the "narrow interactions approach" between organizations is slowly crumbling and this crumbling is not a future-to-come issue but can be observed already. We need technologies to enable effective cross-industry knowledge use that resembles what is occurring to the knowledge instances (*the products*) in the wild and it requires more than the mere exchanging of raw data on the Web.

In this frame, more complex knowledge usage patterns are just one thing but handling more complex knowledge is another. CPS include more and more electronic components and software. In a near future, products will make a stronger use of neuroware and, maybe, organic technology for processing information. This new development and production environment clearly increases the complexity of products, but also increases new types of uses and improves efficiency in resource consumption. Intelligent products have the potential of reducing the amount of resources required to transport, operate and overcompensate eventually higher prices from a reduced production volume with better utility and longer life.

One way to achieve this vision is by harnessing more synergy between deployed products—something that is strongly relying on object's self-knowledge and which can eventually function without *the cloud*. In times of advertising big data this is an often forgotten but a basic and the more tangible motivation behind *Internet of Things:* "the Interacting Things".

Efficient, safe and secure products have the tendency to increase the number of requirements to be satisfied. Large global organizations are believed to be the only way to realize implementation of them because only sufficiently large organizations can employ mass production as means to dilute development cost. However, in a saturated global market, customers can be only convinced by products with better individual fit. This leads to a raise of product variants but the variants' series sizes will continuously drop as they will cannibalize each other. Ultimately, the series size is one—and some industries are close to that point.

Since a radical change towards industry-grade on-demand production of complex individual system variants would raise engineering and production costs by factors of tens to hundreds, we are proposing not to look into some fractional efficiency gains but into major cuts of cost by factors of tens or hundreds. Since so much effort is put into manually organizing storage, transport and retrieval of design data and because automations rely on well-defined access to data, we are proposing to go for a concept to severely cut cost in these areas which is dubbed the *Advanced Industrial Technical Interoperability Concept* (TIC).

What is it about? It is clear that the main cost driver in complex systems development is the inability to provide design knowledge to any part of the engineering process pipeline. This involves a tremendous amount of manual work to find, verify, interpret and transform information about the work products. That is why, the successful avoidance of any manual transport and access to information in the design evolution process is not just a nice to have convenience, it is a game changing event and a massive transformation force on everything related to engineering, production and consumption of outputs as it provides, for the first, interfaces for computerized co-engineering and later fully automatic engineering based on broad cross-domain knowledge sources.

From a practical point of view, it is important to emphasize that many engineering activities are considered "creative" because accessing and transforming pieces of information (knowledge) and turning them into something else is not well accessible for computerized processing—otherwise excellent engineering is quite methodical and yes, even algorithmic. This means that engineering is excellently suited for applying plug-in expert systems. As a rule of thumb, such plug-in components for engineering will only be as good as the input provided to them. Historically, providing knowledge to expert systems has been the main obstacle to their application and hence was inhibiting broad proliferation.

Since providing knowledge as a one-time event is so difficult, the TIC follows the idea of continuous knowledge build-up during all engineering activities. In our eyes, creation of new technology for managing executable or potentially executable forms of knowledge is absolutely feasible with current state of computing and networking technology but requires a new line of development which is totally committed to this goal.

8.3 Related Work

Interoperability has been always a major challenge in the information technology industry [10, 16, 24]. The ability of communicating and exchanging data, information and knowledge between two systems is an active research area in which a quite good number of standards data models, communication protocols, query languages and services can be found applied to domains such as supply chain [2, 4] or the management of industrial processes [3]. Since the complex systems development area is living a digital transformation from document-centric engineering environments to service and data-based environments, interoperability is becoming more and more critical to enable system artifacts reuse and to empower the concept of continuous and collaborative engineering. In this context, existing standards such as the ISO-10303, ReqIF, SysML, UML, etc. have addressed the main needs of exchanging artifacts metadata and contents under standardized data models and protocols. However, the upcoming service environment also needs the promotion of existing file-based approaches to a service-based environment. In this context, recent times have seen the development of the Open Services for Lifecycle Collaboration [29] (OSLC) initiative to tackle some of the existing needs regarding interoperability and integration in the Systems Engineering (SE) discipline.

This emerging effort is seeking new methods to easily integrate SE tools and build collaborative development and operational environments [5]. At present time, OSCL is comprised of several specifications to model shared resources between applications with the aim of sharing more data, boosting the use of web-based standards (Resource Description Framework-RDF and HTTP) and delivering robust products through the collaboration of development and operational tools. OSLC Core [29] is now an OASIS standard, for products and services that support all phases of the software and product lifecycle. Its main objective lies in the integration between work products that support Application Life-cycle Management (ALM) and Product Life-cycle Management (PLM).

More specifically, the Resource Description Framework [7] (RDF), based on a graph model, and the Web Ontology Language (OWL), designed to formalize, model and share domain knowledge, are the two main ingredients to reuse information and data in a knowledge-based realm. Thus, data, information and knowledge can be easily represented, shared, exchanged and linked to other knowledge bases through the use of Uniform Resource Identifiers (URIs), more specifically HTTP-URIs. As a practical view of the Semantic Web, the Linked Data initiative [8] emerged to create this large and distributed database on the Web by reusing existing and standard protocols. In order to reach this major objective the publication of information and data under a common data model (RDF) with a specific formal query language (SPARQL) provides the required building blocks to turn the Web of Documents into a real database or Web of Data and to boost technology-based sectors [11].

OSLC, taking advantage of this new Linked Data environment, is a clear representative of pursuing reuse via universal data formats and relatively standardized software functionality, as could be provided by the OSLC4J framework[2] included with Eclipse Lyo. Similar to OSLC, Agosense Symphony[3] offers an integration platform for application and product lifecycle management, covering all stages and processes in a development lifecycle. It represents a service-based solution with a huge implantation in the industry due to the possibility of connecting existing tools. WSO2[4] is another middleware platform for service-oriented computing based on standards for business process modeling and management. However, it does not offer standard input/output interfaces based on lightweight data models and software architectures such as RDF and REST. Other industry platforms such as PTC Integrity, Siemens Team Center, IBM Jazz Platform or HP PLM are now offering OSLC interfaces for different types of artifacts and development processes.

The other main line of research seems to focus on improving functional deployment which can then operate on specialized data. As it has been previously introduced, service oriented computing [18] offers a new environment to enable the reuse of software in organizations. In general, a service oriented architecture comprises an infrastructure (e.g. Enterprise Service Bus) in which services (e.g. software as web services) are deployed under a certain set of policies. A composite application is then implemented by means of a coordinated collection of invocations (e.g. Business Process Execution Language). In this context, Enterprise Integration Patterns (EAI) [14] have played a key role to ease the collaboration

---

[2]https://wiki.eclipse.org/Lyo/LyoOSLC4J.

[3]http://www.agosense.com/agosense.symphony.

[4]http://wso2.com/.

among services. Furthermore, existing W3C recommendations such as the Web Services Description Language (WSDL) or the Simple Object Access Protocol (SOAP) have improved interoperability through a clear definition of the input/output interface of a service and communication protocol.

In order to improve the capabilities of web services, semantics was also applied to ease some tasks such as discovery, selection, composition, orchestration, grounding and automatic invocation of web services. The Web Services Modeling Ontology (WSMO) [26] represented the main effort to define and to implement semantic web services using formal ontologies. OWL-S (Semantic Markup for Web Services), SA-WSDL (Semantic Annotations for WSDL) or WSDL-S (Web Service Semantics) were other approaches to annotate web services, by merging ontologies and standardizing data models in the web services realm.

However, these semantics-based efforts did not reach the expected outcome of automatically enabling enterprise services collaboration. Formal ontologies were used to model data and logical restrictions that were validated by formal reasoning methods implemented in semantic web reasoners. Although this approach was theoretically very promising, since it included consistency checking or type inference, the reality proved that the supreme effort to create formal ontologies in different domains, to make them interoperable at a semantic level, and to provide functionalities such as data validation, was not efficient. More specifically, it was demonstrated [25] that, in most of cases, data validation, data lifting and data lowering processes were enough to provide an interoperable environment.

In the specific case of software engineering [28], the application of semantics-based technologies has also been focused in the creation of OWL ontologies [9] to support requirements elicitation, to model development processes [17] or to apply the Model Driven Architecture approach [13], to name just a few. These works leverage ontologies to formally design a meta-model and to meet the requirements of knowledge-based development processes. In contrast to these approaches, TIC's Bubble absorbs practical orders provided by tools and the users are left to transform them into a more unified ontology later, when they see fit.

In conclusion, it is clear that system artifacts reuse via interoperability in a service-based environment is an active research area that evolves according to the current trends in development lifecycles including knowledge management, service-oriented computing and micro services. It may have the potential of leveraging new technologies such as the web environment, semantics and Linked Data. However, data exchange does not necessarily imply knowledge management. From service providers to data items, a knowledge strategy is also required to really represent, store and search system artifacts, metadata and contents and enable automatic configuration of development environments.

On the other hand, the need of interoperability in a service-based environment is clear since a quite good number tools have been implemented to support the view of product line engineering and software product lines. It is possible to find tools for production management which main priorities are to control costs and keep the stock up to date. The use of operational data (maintenance and technical data) is a key driver for the proper development of a business strategy. In the hardware development activities, management of standard components and revisions of design are critical. Finally, in software development, the incremental control of design refinement is considered a crucial activity.

A tangled environment of tools has led to the development to various types of tools such as ERP (Enterprise Resource Planning), PDM (Product Data Management), PLM or ALM tools. They are the "spine" of any engineering (or production-based) or business sector but this situation has also implied a lack of interoperability and integration between the different lifecycle development stages (and tools). In the best case, just some pointers between entries in a common databases exist. That is why, in the aforementioned project (CRYSTAL), all stakeholders in the engineering process have attempted to provide a standardized web-based technology (OSLC) for bridging the gap between the different stages of development through a set of traceability links that allow engineers to have a holistic view of the systems development lifecycle and to reuse system artifacts.

In the context of OSLC for product line engineering, each work product contains certain links that are reused from old object revisions or are created to link new objects which already exist. Later, engineers can navigate these links in order to understand how design decisions were made or which quality in the process was achieved. Fig. 8.1 shows this process of iterative linking. The main drawback of this approach is that links tend to point back in time and navigation along those links can become a travel through time. In order to overcome this problem, tools must first point to a "volatile" baseline (a kind of *stream head*) and link target revisions are only frozen after the stream head becomes a frozen baseline (in this point, an engineer is able to navigate consistently within a baseline). The idea behind traceability via web-based links, see Fig. 8.1, is that edited objects are exposed as web resources creating a Linked Data graph. New objects can be linked to existing objects (in their respective revisions) if they are considered relevant as input by the respective roles. Forward linking is also possible either by managing bi-directional links or by providing reverse-lookup with a component like IBM Rational Engineering Lifecycle Manager (RELM).

Fig. 8.1 Linking objects in the development lifecycle using OSLC

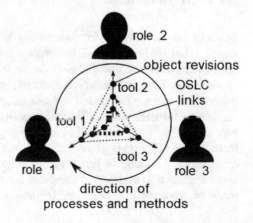

However, there is another issue: the global or absolute linkage (*How to keep links up to date between projects?*). In this case, a careful planning of traceability links and sophisticated link resolution techniques are required. It is possible to have the equivalent of "relative" links. Now, the question is why would one want to have "relative" links in a systems engineering environment in the first place?

Relative links allow "knowledge components" to be reused in various places. For work based on such templates it is important that links point to a place where something *should* be rather than is. *Collecting such dangling pointers* is the equivalent of collecting *TODO*-markers from sources. It pushes the question about the levels of *composability* of document templates, process templates or quality and analysis paths. It also pushes the question which *components* the engineering process is using. Today, we find most of the processes to be a fine grained assembly of elementary steps—very document-oriented, not very structure-oriented. Therefore, for practitioners it seems to be not realizable to reuse greater chunks of related objects from previous projects because they have no proper means to quickly prune irrelevant information and to keep meaningful information. At a document-basis, indeed, it is very difficult to establish. As a consequence, the reuse-rates across projects are rather small and increasing the level of reuse was one of the most desired features when engineers were interrogated [21]. Finally, a previous work on the concept of "BUBBLES" [20] has posed the foundations to provide a sophisticated solution to the problem of absolute and relative links. BUBBLES can theoretically be joined, contain orders and namespaces. This was all considered necessary in order to design a robust computational concept that satisfies sophisticated industry needs and exposes a certain amount of robustness and flexibility in engineering scenarios.

8.4 The Advanced Industrial Technical Interoperability Concept (TIC)

8.4.1 Introduction

Building on the need of interoperability and link management, a new question arises: *What addresses shall be relative and whether (if provided) such reused components can be combined with the necessities posed by the project?*

As it has been previously introduced, today's product integrations are weak integrations and are advertised as loose couplings. However, moving towards new fascinating business cases will require more complex, more parametric interactions between all products involved in engineering (this of course also implies a proper inclusion of all "non-engineering" tools such as word processors or spreadsheet applications). Linking is just a contributing factor to manage complexity. Links allow engineers to understand structures but for that purpose the right kinds of links must be generated a priori to (planned) evaluations. If necessity of certain evaluations comes up after a problem was detected then the question that arises is how existing link information can contribute to answering questions behind these new evaluations.

It has been found that the creation of links to provide certain conceptual derivations (*traceability links*) is not enough for managing complexity. There are many cross-product or even cross-project questions that require other types of links that are usually manually created. Since many engineering environments do not consider full scope of links, a lot of linking potential is unused or is later not usable in order to answer the actual questions. Some engineers could even consider them pure audit satisfaction. The worst of all is that creating links is a time consuming task. It is crucial to design new technology to reduce the amount of linking to a minimum. Any link that is created must be *important*, i.e. engineers are spending a high-quality time thinking about it, about its role in the engineering and production environment. This work must not be quickly invalidated and there must be high yields from it, measurably from a business and a personal point of view. Therefore, link reuse and automatic link management as far as this is possible, is a central theme in designing new interoperability technology. It must not only connect data sources to sinks but must help to manage complexity in data.

Since the efficiency in linking work products deeply relies on intended analysis which is often omitted, engineers look for other technical means to answer their questions once they pop up. For instance, *Query interfaces to virtual databases* seem to offer the necessary functionality. Setting up such a virtual database has many degrees of freedom. For example the Porsche spin-off *Softwarehelden* has implemented a virtual database platform *Cluu*[5] for relational databases. The queries run against that virtual databases which can connect various relational sources of information. However, updating databases via queries is a generally tricky issue that requires exact understanding of all administrative entries in a database. Therefore, many integrators following a query-based approach will look for SPARQL/SPARUL interfaces.

Such web-based query interfaces allow engineers the interrogation of many databases but they have to translate their internal models to RDF for that purpose. The amount of effort to provide a purely logical view of the data (i.e. without all technical hassles necessary to model all application-specific features) as RDF or proprietary representations should not be underestimated. In practice, tool vendors provide proprietary APIs to talk to the objects in the databases without having to know brittle details. These APIs are often better suited to queries than native relational database contents. Because tool vendors shy implementing all too abstract concepts of their data (because they are associated with functional losses) and (to be fair) do not have the knowledge of the needs of many other products, we are facing the situation that most software tools relying on queries as means of integration are bound to a specific product group using particular query APIs.

---

[5]http://softwarehelden.com/.

However, the consequence of this situation is that such integration technologies are *developer technologies* and not *engineer technologies*. All companies desire technology for their tools which would allow their users to quickly access, reconfigure and maintain applications without programming skills (and more importantly *without special development environments*). Queries do not naturally bring all the features demanded by user-scenarios but some products bring features of this kind such as IBM's Change Synergy combo. Satisfying such scenarios on a general scale would require many free-to-use standard tooling components which are sufficiently useful and sufficiently easy to use by users (who are systems engineers).

It is clear that there is product complexity and complexity of the informational entanglement between engineering artifacts that is not well understood as linking or querying problem. For instance, in the CRYSTAL project, the actual mission was to achieve maximum levels of "single source of truth" by keeping the data in one tool and not to copy data anywhere. However, in the engineering process, many pieces of information are *entangled*. Such entanglement is either trivial like for example a duplicate but more complex entanglements exist such as that one variable cannot exceed a certain range of values if a different variable is in a certain state. E.g. a test specification will rely on a specific value encoded into a requirement. This makes these two objects entangled.

In order to evade synchronization, the data concepts must be strongly normalized as is well understood for databases. Normalization of data concepts has the tendency to extract informational entanglement of information objects by *atomizing* them. In the case of our example, it seems futile to hope that natural language requirements and test specifications will be ever perfectly normalized. This is true for many practical examples. Accepting the existence of basic practical limits to normalization of engineering data pulls up the question of various forms of data synchronization (see Fig. 8.2).

In fact, even in the CRYSTAL project, the dominating number of use case solutions was relying on some form of data synchronization. Even if OSLC provides a standardized representation of certain data types as RDF, it is not intended for sophisticated data synchronization which shall be avoided by OSLC philosophy. Customers looking for sophisticated synchronization mechanisms will look for products such as Agosense Symphony or Knowledge Inside ArKItect.[6]

These import-export-suites concentrate on a broad coverage of products. The adapter framework mainly concentrates on emulating a neutral communication channel and to provide user tangible management tools for exercising a programming-free control over interoperability cases. On the contrary, projection databases' main approach is to emulate a neutral data storage and to support various synchronization operations on them. The interoperability is established by implementing round-trips and projections from one database to another. The foundations of the TIC layer are a mix of both approaches. On one hand it acts as a neutral

---

[6]http://www.k-inside.com/web/.

Fig. 8.2 Terminology usually associated with synchronization of data

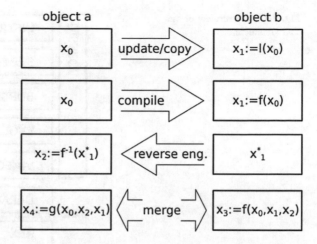

communication channel between sinks and sources. On the other hand, it has also the ability to technically buffer and implicitly model an intermediate database which appears to consumers in "the way they like it".

8.4.2 Description of the TIC Layer

Since the mission of the TIC is not mainly about interoperability but about making knowledge iteratively automatically processable, the TIC outgrows other interoperability technologies by complexity, see Fig. 8.3. This architecture is, on the one hand, sufficiently general purpose in order to answer almost all necessary interoperability use cases found in industry but, on the other hand, is excellently suited to rationalizing design materialization activities.

The TIC architecture, see Fig. 8.3, consists of several conceptual layers such as the communications layer, in it the *AEX* (the optional Advanced Exchange Protocol), *DERAIL* (not an acronym), the *VIMECS* (Viral Messaging and Call-back System), the *EVOL* (the Extensible Virtual Object Layer), *BUBBLES* (not an acronym), *INSTRUMENTS* (not an acronym), the *CAGUIDS* (Cascading Globally Unique Identifier System) and a collection of several additional, engineering and production motivated concepts subsumed under the term *DEA* (Development Environment Architecture). This platform looks for compiling together all perspectives for the future development of industrial products keeping the capability of supporting sophisticated and innovative industry scenarios. Thus, once a technical infrastructure is developed based on the presumed technical interoperability concept, the implementation effort is considered finite—which is a critical condition to assessing the relationship between cost and benefit.

The main foundation behind the TIC is to abandon the idea of incrementally integrating industry sector by industry sector ("silo by silo"). Such process is

Fig. 8.3 An overview of the
TIC architecture

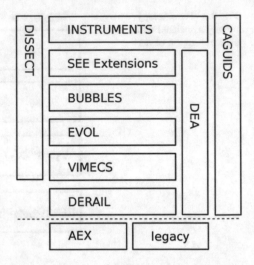

unnecessarily slow and danger since this strategy will eventually never achieve the full potential of digitization of industries because killer applications are not in view—in such process they would only pop up accidentally. We propose a more consciously designed grand technology with certain killer applications on mind such as:

- The never ending series.
- Consumer co-design of product.
- Pro-active production facilities design.
- Production facility as a product.
- Rapid product design on demand.
- New levels of knowledge transfer between science and industry in terms of mass and speed.
- Late, dynamic safety validation and safety generation for new and unprecedented product assemblies.
- Decentralized, close to product owner storage of data (better privacy satisfaction, fewer IT resource necessary).
- Easy product-on-product business models.
- …

From this perspective, the TIC is designed to overcome all industry silos, see Fig. 8.4, by creating a global platform with universally available new computing concepts from the TIC. This platform shall span mostly all areas of engineering, production, product deployment, phase of use and product afterlife to the concept of *fabrication web*). The goal of this *fabrication web* is to absorb design and production intelligence during engineering, production and usage activities and to highly automate its further transport and consumption until deep into other environments, see Fig. 8.5. The knowledge transportation process shall be omni-directional, i.e. it is not designed around information exchange patterns which

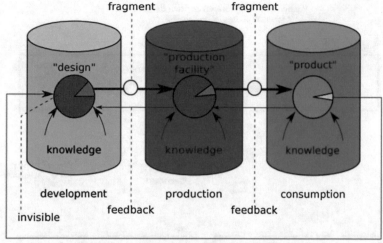

Fig. 8.4 An example of industry silos in which each silo is transforming the product having a partial view of other silos

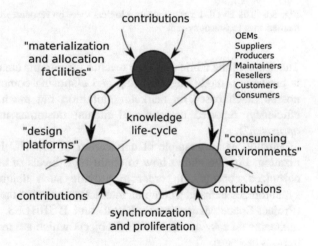

Fig. 8.5 An example of the TIC applied to silos without having any limited scope for any design, control or status information in the industry

are currently dominating. This knowledge absorption is only desirable if it can be controlled (security) and if, ultimately, it can be turned into executable knowledge (automation).

Figuratively speaking, the fabrication web shall be like a sponge soaking in technologically transformable knowledge until becoming an expert system on its own in more and more areas of design, production, deployment and decommissioning. This knowledge shall also become part of the products (self-aware products) which can contribute to safe and innovative uses in unprecedented situations. This leads to reduction of human involvement over time in any part of the industry.

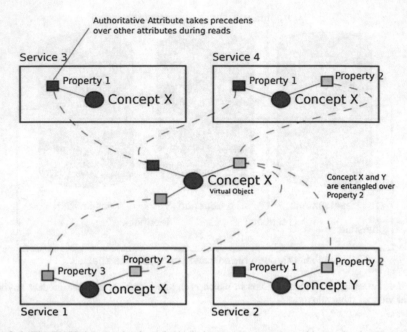

Fig. 8.6 The EVOL layer maintains a logical view on resources stored in a redundantly scattered manner in a technical system

The scope of the fabrication web reaches far into the area of product application and is producing, modifying, updating and disposing components of the environment not so much based on marketing intuition but much more on combination of efficiency, reported demand and factual customer use ("*resonance-theoretical approach*").

Finally, as an example of application of the TIC layer for keeping synchronization, Fig. 8.6 shows how to attain high levels of consistency relying on synchronized concepts. In order to provide such ubiquitous capability, the TIC synchronizes items in a complex interaction between the layers, mainly DEA: PSM (Project Space Management), EVOL and BUBBLES. At minimum, this system maintains an easy access to virtual objects which are redundantly scattered around various storages.

8.4.3 Basic Interoperability Functions

The basic function of an interoperability layer is to transport information from various sources to sinks. In this process, various kinds of incompatibilities can occur. We have categorized eight types of basic forms of incompatibility which all must be addressed in order to attain strong forms of interoperability. Thus, we can find: Carrier, Size, Semantic, Variability, Expressiveness, Temporal Deadline, Reliability and Dynamical incompatibilities.

Unlike most other interoperability technologies, the TIC does not attempt to solve compatibility in a global frame. The TIC manages interoperability at a meta-level: a key idea in the TIC is that communications are bound to communication frames in which compatibility must be established (or incompatibilities of above kinds avoided). Therefore, it is not necessary to find a piece of executive technology which has to always satisfy a global set of requirements which would have relatively little chance to be satisfied. For example, one tool can only be included via web interfaces and another tool can only consume files. The TIC has concepts how this broad range of storage and transport technology can be treated uniformly, so that essential project conventions are maintained.

The communication frames are explicit in the TIC. Communication partners attach to such a frame and can perform several basic functions:

- Detect interfaces required for a toolchain to work.
- Measure compatibility of interfaces in regard to the communication frame.
- Install and configure new components in order to establish compatible interfaces for a given communication frame.
- Report on attained standards and bottlenecks (e.g. manual interfacing) centrally to the authority of the context.

8.4.4   Application of the TIC Layer to Manage Complex Design Reuse Scenarios

During the conceptual work on the TIC, we have considered reuse not only at the product design level but also at the process reuse level: the interoperability layer shall not only simplify the reuse of work on design but also foster reuse of reproduction and qualification work. The basic idea behind reusing product design has been motivated and described in more detail in [20].

Since reuse is never without any change to reused entities, it was a central goal to reduce the amount of effort related to practicing reuse. Currently, too often, developing new things is easier and less error prone than trying to reuse results. This can only be changed if technology can natively support reuse practice and this support must include management of reused components, their clear visibility even after the development continued for some while, reuse of links, instantiation of templates to instances and communications regarding reused objects across projects.

Current state of the art solutions, such as streams and *change set forwarding* mechanisms are only partial solutions to the reuse problem. Best products implement such features for themselves and therefore, effectiveness of such mechanisms are quickly limited if products from other vendors are mixed or blended into the toolchain. We therefore concluded that management of reuse practice is mandatory requirement for any industrial interoperability layer.

On the contrary, the TIC is designed to provide high levels of reuse support even if the tools have hardly any support for the necessary mechanisms. However, during systematic evaluations [20] we have found out that reuse requires even finer control than provided with Bubbles alone.

In order to support desired development practices we have introduced a rectangular default organization of the Bubble which was simply named the GRID. The GRID and methodologies around it replace the V-style engineering. The GRID is much better suited to what engineering companies actually do than the *Vee-model* and we are convinced that GRID methodology guarantees excellent work product coherence between collaborating organizations and sophisticated toolchains-something that is not always true with V-style methodologies.

In future work, we will publicly lay out the details of GRID and reuse with GRIDed Bubbles as these concepts are fresh CRYSTAL project output. An example of what can be done with GRIDed Bubbles, see Fig. 8.7. In particular, it shows how a design (held by a Bubble) is aware of various degrees of its refinement (CCS layers), domain aspects (A) or representation modes (TRL or Readiness). When duplicating this design a limiting factor can be defined which tells what kind of forwarding mapping shall be practiced between original and reused Bubble. With such capabilities it is very easy to organize perfectly managed first design explorations, design family refactoring or design drift corrections.

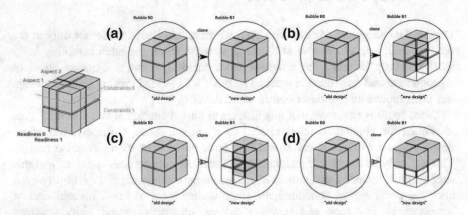

Fig. 8.7 Cloned or derived Bubbles can limit the scope of inheritance along GRID coordinates. **a** Full scope inheritance, **b** inherit only early low readiness design, **c** inherit certain aspect knowledge, **d** inherit only abstract design knowledge

8.5   Discussion of the TIC Layer in Industry 4.0 Toolchains

In this section, we argue against the conventional design of interoperability technologies for narrow domains, as we are observing that strongly overlapping sets of concepts are used for them. Unfortunately, this does not make them any close to compatible. Each such technology and related product families are posing distinct media for transport and storage of data. Using data across such media is almost always accompanied with losses of management quality or confidence in data fidelity. From such experience, we urge industrial and scientific communities to define and implement a *free to use* and openly specified full spectrum interoperability layer. In- vestments into a new technology based on concepts of this kind can only be justified if a credible argument can be made that:

1. New type of use-cases or businesses are possible which are not possible with fragmented interoperability technologies.
2. New levels of cost efficiency can be attained (reduces more cost than it adds).
3. Is indeed pragmatic enough for all users participating in exchanges.
4. Demonstrates sufficient robustness.

Interoperability technology which could encompass several industrial domains must support major functions along several dimensions. This will certainly result in a complex technology which will require a concept or architecture in order to satisfy the many requirements flowing into its design. Among the most desired properties of an interoperability layer are:

- Easy to use by the user.
- Must offer administration features on industrial scale.
- Must allow collaboration on projects with hybrid toolchains.
- Must allow collaboration on projects between different companies.
- Must gracefully accept and integrate non-TCP/IP communications.
- Must support exact traceability of engineering and other processing activities.
- Must be able to extend from development department until the end consumer.
- Must support any type of organization of data.
- Must support activities on real objects (not only digital ones).
- Must support extreme re-use scenarios.
- Must be feasible with current technological resources.
- ...

Furthermore, the following list of high level technical features should be expected from a sophisticated new interoperability layer based on the TIC:

- Transport data between places, times and tools.
- Transport data between projects and their governing conventions.
- Transport data between organizations.
- Transport events, structured data, complex data and stream data.

- Transport data under general constraints, such as real-time requirements.
- Transport of data with managed data transformation.
- Allow piecewise transports on terms of infrastructure.
- Allow reuse of digital entities like with real entities.
- Virtualize all high level data management in order to allow slow tools landscape evolution.

Since interoperability is not only concerned with easing transport of data, but also with the selection of the presentation of data, its reuse and organized processing, the relationship between data and knowledge is appalling. We therefore provided a synthetic concept at the heart of the TIC that is something between a context, data container, signal routing facility and (in extended sense) a variability management tool. This concept is the *Bubble*. As we have found out, Bubbles are not only a useful concept for managing related pieces of data between tools but also to organize production and use of physical collections. For example, it is theoretically possible to create a complete laboratory model, make it accessible via Bubbles and then to instantiate the respective class Bubble to an instance Bubble which can be replicated in real life (resulting in replication of laboratory). From industrial point of view, and in any meaningful consumer point of view, design is only as valuable as means to instantiate[7] it. This implied for the TIC that design is strictly understood as contributing part to a production facility. The TIC concentrates on a "factory" concept and each bubble is a natural point of production.

Since interoperability is not only concerned with easing transport of data, but also with the selection of the presentation of data, its reuse and organized processing, the relationship between data and knowledge is appalling. We therefore provided a synthetic concept at the heart of the TIC that is something between a context, data container, signal routing facility and (in extended sense) a variability management tool. This concept is the *Bubble*. As we have found out, Bubbles are not only a useful concept for managing related pieces of data between tools but also to organize production and use of physical collections. For example, it is theoretically possible to create a complete laboratory model, make it accessible via Bubbles and then to instantiate the respective class Bubble to an instance Bubble which can be replicated in real life (resulting in replication of laboratory). From industrial point of view, and in any meaningful consumer point of view, design is only as valuable as means to instantiate it (*instantiation* means any kind of virtual or physical production or reproduction). This implied for the TIC that design is strictly understood as contributing part to a production facility. The TIC concentrates on a "factory" concept and each Bubble is a natural point of production.

---

[7]We are using the term instantiation and mean by it any kind of virtual or physical production or reproduction.

8.6 Conclusions and Future Work

The coordination of the future industry and the prevention of cost explosion will require new degrees of digital interaction, supra-organizational planning and on-demand distribution. As real examples of similar approaches, the German initiative *Industry 4.0* and the *Industrial Internet* movement in America have set out to provide technology for (almost) this purpose. However, parties involved in these movements face the simple fact that achieving interoperability between design and production facilities fails for even simple types of data, such as requirements, change requests or status information. The general question is then why the most basic things simply do not work in terms of interoperability.

To our convincing, the problem of today's, lack of interoperability, lies not in the technical difficulty to provide interoperability but the lacking understanding about how businesses can survive and can keep making business after buying deep into open interoperability. Currently, "just connecting" and "just automating" the industry is not attractive for many companies as they feel threatened to be degraded to bigger companies' departments. Moreover, buying in on specific vendors' software in order to achieve necessary automation and interoperability is making companies vulnerable to monopolies. As a consequence, there is no motivation to invest into communications compatibility and richer communications as the benefits of these are rather hazy.

Nevertheless, the first companies who are successfully employing and understanding far reaching implications of new interoperability technology and who can handle sophisticated new business cases with it will put all other industries under significant pressure to survive.

Acknowledgements The research leading to these results has received funding from the ARTEMIS Joint Undertaking under grant agreement N 332830-CRYSTAL (CRitical sYSTem engineering AcceLeration project) and from specific national programs and/or funding authorities. This work has been supported by the Spanish Ministry of Industry.

References

1. AG, D.: Industry 4.0: Challenges and solutions for the digital transformation and use of exponential technologies. Consultancy report, Deloitte AG (2015). https://www2.deloitte.com/ch/en/pages/manufacturing/articles/manufacturing-study-industry-4.html
2. Alor-Hernández, G., Sánchez-Ramírez, C., Cortes-Robles, G., Rodríguez-González, A., García-Alcaráz, J.L., Cedillo-Campos, M.G.: BROSEMWEB: a brokerage service for e-Procurement using Semantic Web Technologies. Comput. Ind. **65**(5), 828–840 (2014). doi:10.1016/j.compind.2013.12.007. http://linkinghub.elsevier.com/retrieve/pii/S0166636151 3002492
3. Alor-Hernández, G., Sánchez-Ramírez, C., García-Alcaráz, J.L. (eds.): Handbook of Research on Managerial Strategies for Achieving Optimal Performance in Industrial Processes. IGI Global (2016). http://services.igi-global.com/resolvedoi/resolve.aspx?doi=10.4018/978-1-5225-0130-5

4. Alvarez-Rodríguez, J.M., Labra-Gayo, J.E., de Pablos, P.O.: New trends on E-Procurement applying semantic technologies: current status and future challenges. Comput. Ind. **65**(5), 800–820 (2014). doi:10.1016/j.compind.2014.04.005. http://www.sciencedirect.com/science/article/pii/S0166361514000803
5. Alvarez-Rodríguez, J.M., Llorens, J., Alejandres, M., Fuentes, J.: OSLC-KM: a knowledge management specification for OSLC-based resources. INCOSE Int. Symp. **25**(1), 16–34 (2015). doi:10.1002/j.2334-5837.2015.00046.x
6. Austin, S.: What is the Fourth Industrial Revolution? (2016). http://blog.obrary.com/what-is-the-fourth-industrial-revolution
7. Beckett, D.: RDF/XML Syntax Specification (Revised). W3c Recommendation, W3C (2008). http://www.w3.org/TR/rdf-syntax-grammar/
8. Bizer, C.: The Emerging Web of Linked Data. IEEE Intell. Syst. **24**(5), 87–92 (2009). doi:10.1109/MIS.2009.102. http://ieeexplore.ieee.org/lpdocs/epic03/wrapper.htm?arnumber=5286174
9. Castañeda, V., Ballejos, L., Caliusco, L., Galli, R.: The use of ontologies in requirements engineering. Glob. J. Res. Eng. **10**(6) (2010). http://engineeringresearch.org/index.php/GJRE/article/view/76
10. Clark, T., Jones, R.: Organisational interoperability maturity model for C2. In: Proceedings of the 1999 Command and Control Research and Technology Symposium (1999)
11. Colombo-Palacios, R., Sánchez-Cervantes, J.L., Alor-Hernández, G., Rodríguez-González, A.: Linked data: perspectives for IT professionals. Int. J. Hum. Capital Inf. Technol. Professionals **3**(3), 1–12 (2012). doi:10.4018/jhcitp.2012070101. http://services.igi-global.com/resolvedoi/resolve.aspx?doi=10.4018/jhcitp.2012070101
12. Consulting, C.: Industry 4.0—The Capgemini Consulting View. Consultancy report, Capgemini (2015). https://www.de.capgemini-consulting.com/resource-file-access/resource/pdf/capgemini-consulting-industrie-4.00.pdf
13. Gaevic, D., Devedic, V., Djuric, D., SpringerLink (Online service): Model Driven Architecture and Ontology Development. Springer, Berlin (2006). http://proxy.library.carleton.ca/login?url=http://dx.doi.org/10.1007/3-540-32182-9
14. Hohpe, G., Woolf, B.: Enterprise Integration Patterns: Designing, Building, and Deploying Messaging Solutions. The Addison-Wesley Signature Series. Addison-Wesley, Boston (2004)
15. Exman, Iaakov, Fraga, Anabel, Llorens, Juan, Alvarez-Rodríguez, Jose Maria: SKY-Ware: the unavoidable convergence of Software towards Runnable Knowledge. J. UCS (JUCS) **21** (11), 1405–1424 (2015)
16. Kasunic, M.: Measuring Systems Interoperability. Software Engineering Institute. Technical report (2003)
17. Kossmann, M., Wong, R., Odeh, M., Gillies, A.: Ontology-driven Requirements Engineering: Building the OntoREM Meta Model, pp. 1–6. IEEE (2008). doi:10.1109/ICTTA.2008.4530315. http://ieeexplore.ieee.org/lpdocs/epic03/wrapper.htm?arnumber=4530315
18. Krafzig, D., Banke, K., Slama, D.: Enterprise SOA: Service-Oriented Architecture Best Practices. Prentice Hall Professional (2005)
19. Lee, E.A.: Cyber physical systems: Design challenges. In: 2008 11th IEEE International Symposium on Object and Component-Oriented Real-Time Distributed Computing (ISORC), pp. 363–369. IEEE (2008)
20. Lodwich, A., Álvarez-Rodríguez, J.M.: Bubbles: a data management approach to create an advanced industrial interoperability layer for critical systems development applying reuse techniques. CoRR abs/1605.07336 (2016). http://arxiv.org/abs/1605.07336
21. Marko, N., Liebel, G., Sauter, D., Lodwich, A., Tichy, M., Leitner, A., Hansson, J.: Model-based engineering for embedded systems in practice (2014)
22. Nanterme, P., Daugherty, P.: People First: The Primacy of People in a Digital Age. Technical report, Accenture Inc. (2016)
23. Pohl, K., Bӧckle, G., van Der Linden, F.J.: Software Product Line Engineering: Foundations, Principles and Techniques. Springer Science & Business Media (2005)

24. Rezaei, R., Chiew, T.K., Lee, S.P., Shams Aliee, Z.: Interoperability evaluation models: a systematic review. Comput. Ind. **65**(1), 1–23 (2014). doi:10.1016/j.compind.2013.09.001. http://dx.doi.org/10.1016/j.compind.2013.09.001
25. Rodríguez, M.G., Rodríguez, J.M.A, Muñoz, D.B., Paredes, L.P., Gayo, J.E.L., de Pablos, P. O.: Towards a practical solution for data grounding in a semantic web services environment. J. UCS **18**(11), 1576–1597 (2012). 10.3217/jucs-018-11-1576. http://dx.doi.org/10.3217/jucs-018-11-1576
26. Roman, D., Keller, U., Lausen, H., de Bruijn, J., Lara, R., Stollberg, M., Polleres, A., Feier, C., Bussler, C., Fensel, D.: Web service modeling ontology. Appl. Ontol. **1**(1), 77–106 (2005)
27. Schwab, K.: World Economic Forum: The Fourth Industrial Revolution (2016)
28. Valencia-Garcia, R., Alor-Hernández, G.: Special issue on knowledge-based software engineering. Sci. Comput. Program. **121**, 1–2 (2016). doi:10.1016/j.scico.2016.02.005. http://linkinghub.elsevier.com/retrieve/pii/S0167642316000496
29. Workgroup, O.C.S.: OSLC Core specification version 2.0. Oasis Standard, OASIS (2015)

Part III
Knowledge-Based Decision Support Systems (Tools for Industrial Knowledge Management)

Part III
Knowledge-Based Decision Support
Systems/Tools for Industrial
Knowledge Management

Chapter 9
Knowledge-Based Decision Support Systems for Personalized *u*-lifecare Big Data Services

Muhammad Fahim and Thar Baker

Abstract The emergence of information and communications technology (ICT) and rise in living standards necessitate knowledge-based decision support systems that provide services anytime and anywhere with low cost. These services assist individuals for making right decisions regarding lifestyle choices (e.g., dietary choices, stretching after workout, transportation choices), which may have a significant impact on their future health implications that may lead to medical complications and end up with a chronic disease. In other words, the knowledge-based services help individuals to make a personal and conscious decision to perform behaviour that may increase or decrease the risk of injury or disease. The main aim of this chapter is to provide personalized ubiquitous lifecare (*u*-lifecare) services based on users' generated big data. We propose a platform to acquire knowledge from diverse data sources and briefly explain the potential underlying technology tools. We also present a case study to show the interaction among the platform components and personalized services to individuals.

Keywords Big Data Services · Personalized u-lifecare · Decision support system · Knowledge-based system

9.1 Introduction

We live in the age of data; everything surrounding us is linked to a data source (e.g., smartphone, smartwatch, wearable computing devices including smart glasses) that captures massive amount of valuable data digitally. This data can be utilized to

M. Fahim (✉)
Department of Computer Engineering, Faculty of Engineering and Natural Sciences, Istanbul Sabahattin Zaim University, 34303 Istanbul, Turkey
e-mail: muhammad.fahim@izu.edu.tr

T. Baker
Department of Computer Science, Faculty of Engineering and Technology, Liverpool John Moores University, Liverpool, UK

© Springer International Publishing AG 2017
G. Alor-Hernández and R. Valencia-García (eds.), *Current Trends on Knowledge-Based Systems*, Intelligent Systems Reference Library 120, DOI 10.1007/978-3-319-51905-0_9

187

generate recommendations, monitor physical activities, and generate real-time alerts for different application domains such as sports, lifestyle, and healthcare. The same data is used with new datum to extract hidden trends, relationships, and association [1]. Processing data to extract knowledge is associated with many challenges due to the big data volume, speed through which it is generated, and redundancy and noise [2]. More importantly, we lack the time and capacity to study and search the collected data manually in real time in order to build knowledge that helps in steering our future. These challenges signify the need to the development of new technologies, tools, and practices for collecting, integrating, analyzing, and presenting a large volume of information to enable better decision making [3].

Consider a daily life routine scenario, where human interacts with multiple smart devices and requests personalized services for daily routines, softening the mood, healthy food or exercise routines. If we are able to process the data around human, and construct and manage the knowledge over the reliable IT infrastructure, then providing the personalized u-lifecare services at the right time becomes possible.

Single source of information is neither adequate nor sufficient to understand human, as human cognition comes from the knowledge of physical, mental, and social contexts. Furthermore, the behavior changes also have high impact to understand daily routines. Currently, the main challenge that remains under investigated is how to acquire knowledge from diverse data sources in efficient time and cost. In order to process such big data, acquire knowledge and build decision support system, we propose a platform that will accommodate diverse sources of structured and unstructured data and briefly explain the underlying technologies and tools to process the produced big data in a cost/time-effective way.

9.2 Related Work

Knowledge-based decision support systems play pivotal role in improving the quality of life by providing tools and services needed to resolve emergent problems, and knowledge necessary to suggest a new/specific strategy to work with various situations [4]. Many researchers and companies focused on exploiting the data for providing human-centric services, but their attempts are still limited. For instance, Nike+ [5], Samsung Gear [6], LG Smartwatch [7], Microsoft Band [8], Fitbit Blaze [9], to name but a few, promote active lifestyle and provide some basic health recommendations based on the recognized human activities, burned calories or hours of sleep. The same data can be used to extract hidden trends, relationships and association with wellness application, chronic disease prevention as well as analyzing the specific group of people.

Bilal et. al. [10] introduced data curation framework to accumulate users' sensory data from multimodal data sources in real time and preserve as a lifelog. They provide management support for the collection of large volume of sensory data. This data will be further processed under the Mining Minds (MM) platform [11] to generate knowledge and recommendation services for individuals.

Reza et al. [12] considered the smartphone as a portable computer. Such devices can be used for personal data collection because users carry their smartphone all the time. Authors also acknowledge the fact that digital data grow rapidly; therefore, it is difficult to process using traditional data management tools and techniques. Hence, Cloud Computing infrastructure along with big data processing provides more opportunities for critical infrastructure systems including health and human welfare, commerce and business, and economic systems, etc. with reduced cost [13, 14]. Miguel et al. [15] introduced a semantic platform for cloud services annotation and retrieval from their descriptions. The system can automatically annotate different cloud services from their natural language description.

The discovery of the knowledge typically involves the use of human expert knowledge and other source of information, which can be stored in logical structures, accessible and readable by machines [16]. Marco et al. [17] come up with semantic representation in terms of ontology-based knowledge to support some phases of the decision making process. The proposed approach has been successfully implemented and exploited in a decision support system for personalized environmental information. Similarly, another prospective of generating knowledge is introduced by Ling et al. [16]. They proposed knowledge generation by assisting the expert-driven rules via a hybrid method, which combines human domain expertise with machine learning methods to provide more accurate, effective and efficient method for discovering knowledge in complex domains.

In order to present the knowledge in a useful manner, there exist numerous challenges. Kambatla et al. [12] highlights those challenges in details to provide useful analytics. In terms of enhancing the wellbeing, data in healthcare poses some of the most challenging problems to large-scale integration as generated data is huge and contains verities including electronic medical records (EMR) and electronic health records (HER), imaging data, or personalized drug response. Similarly, several analysis tasks are also time-critical. For example, patient diagnoses, progression of outbreaks, etc., all have tight performance requirements. So, in this regard, a platform is required that can process such a big data and response in time. Our proposed platform keeps these challenges in mind and carefully takes care of all discussed issues in this section.

9.3 Proposed Platform

The proposed platform is based on big data storage and processing framework, data management, learning models, construction of knowledge bases, and personalized *u*-lifecare services. It is illustrated in Fig. 9.1, and details are given in the following sections.

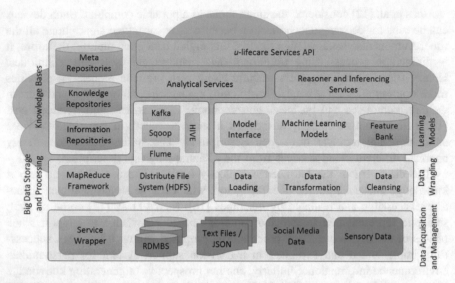

Fig. 9.1 The proposed knowledge-based decision support system for personalized *u*-lifecare big data services

9.3.1 Data Acquisition and Management

According to IBM, 90% of data in the world today has been created only in the last two years [18]. This data comes from heterogeneous data sources and have a lot of varieties including structured, unstructured and partially structured data, as shown in the bottom layer in Fig. 9.1. These data sources consist of multimodal physical sensors, which are based on embedded and different wearable devices such as smartphones and smart shirts [19–21]. It also includes the social networks data that is considered as an input source to our proposed platform. The input is raw data gathered and partially structured with respect to sensor or source categorization in a (csv, xml, dat, JSON, relational data, web scrapping, or text files, etc.). Each data source has its own configuration properties for establishing the connection and grabbing the existing or new data. We designed the data source "*service wrapper*" to collect and store all the technical information that is required to access the data from a particular source as well as data synchronization information. Once we are connected to a data source, we can gather the data and store the logs into Hadoop Distribute File System (HDFS) in real time.

9.3.2 Data Wrangling

One of the major issues encountered when building knowledge-based decision support systems is acquiring high quality of data and reshaping it for further process. In this regard, data wrangling technique shapes and manipulates the raw data to an alternative format that is suitable for exploration and analysis [22]. In other words, it is able to convert and map data from raw form into a useful form that is more convenient for specific use. It includes cleansing, transformation and loading into target repository. Our consideration toward the data generation is big data that means generated data have high volume, velocity, variety and veracity. In this case, very well know technique so-called ETL (Extraction, Transformation and Loading) and models are not suitable because they require manual work from technical and domain experts at different stages of the process [22]. Data wrangling/munging is one the technique that works for both internal and external data sources with semi-automated tools and less human intervention. Details of data wrangling sub-modules namely (Data Cleansing, Data Transformation and Data Loading) are given below.

9.3.2.1 Data Cleansing

The gathered data from a single source, or multiple sources, is deemed correct but contains many inconsistencies and errors [23]. These issues arise because data is not collected for constructing the knowledge base and analysis purposes. In this case, we need to tackle the main issues (i.e., ambiguous or invalid data), missing values, duplicate entries, upper versus. lower case, date and time zones, etc. Treating these issues will increase the quality of data and assist in extracting the appropriate information and facts that lead to the effective decision-making. In order to get the unified format of data, we need to prepare the data in a consistent way and solve the above-mentioned main issues associated with the collected data. In Table 9.1, we

Table 9.1 Solutions to the issues along with illustrative examples

Issues	Examples	Unified format
Inconsistent data	Gender can be M/F or 1/0 or male/female	We need to define the rule, which one should we keep for the further processing of data
Ambiguous or invalid data	Last updated: 2026	This data is considered as invalid, and we can find outliers by visualizing the histogram
Missing values	Sensor reading: _____	We can estimate the missing attribute value based on other examples for which this attribute has a known value Simple way to assign the most common value for that attribute or delete the instance depending on the situation. A more complex technique is to assign a probability to each of the possible values of attributes [24]
Duplicate Entries	User ID field in sensor logs and social media	If we merge these data files then we can keep the single User ID instead of two
Upper versus. lower case	Sitting or sitting	We can define rules for handling case-sensitive matters as well
Date format and time zones	Date: 7/29/2016 13:15 Date: 29/7/2016 12:15 pm	We can define rule for consistent date format and time zone

provide the possible solutions to the issues along with examples for the transformation of data into a unified format.

We can use some existing tools for data cleansing to ease our job such as Data Wrangler [25], Tabula [26], OpenRefine [27], Python and Pandas [28], to name a few.

9.3.2.2 Data Transformation

In data transformation, we will apply the smoothing technique in order to remove the noise and short-term irregularities hence improve accuracy of forecasts and readings. Smoothing also have more positive effects on important processes in data transformation such as pre-joins given that joins are expensive and warrant special performance consideration as they may create data sets beyond what it is needed for so pre-joining data once and store them for further future use deemed requisite; normalization to scale the data in the specified range, discretization by dividing the range of continuous attribute into intervals, and new attribute construction from the given one. The following table highlights the data transformation requisites and corresponding formulas and explanation (Table 9.2).

9.3.2.3 Data Loading

After applying cleansing and transformation techniques, we need to load this structured and partially structured data for the efficient access whenever requested by the other components to process this data. The temporal data is stored in HDFS by utilizing a distributed "Apache Flume" data service [29]. Flume consists of agents, which includes a source, channel, and sink; which all work together to flow data from data sources to the required destination. HDFS has many advantages over the others as open source, scalable, reliable, manageable and moving large amounts of log data, amongst others.

Table 9.2 Transformation formulas

Transformation	Formula/Method	Explanation
Normalization	$Z_i = \frac{x_i - \min(x)}{\max(x) - \min(x)}$	Where $x = (x_1, \ldots, x_n)$ and Z_i is ith normalized data between the range of 0 and 1
Discretization [26]	Threshold method Multi-interval method Gain Ratio method	Discretizing the continuous values attribute into multiple intervals
New attribute construction	Mean, standard deviation, Variance	Statistically attributes help to know more about the data

9.3.3   Big Data Storage and Processing

Data comes from the heterogeneous sources including sensors (i.e., embedded in smartphone, smartwatch or wearable devices), social networks, publically available datasets, and historical data for extracting the valuable information. Consequently, the size, velocity, variety and veracity of the data, is huge and hence requires reasonable time and cost for the processing. Thus, our proposed data storage is based on an open source framework Apache Hadoop that supports data intensive jobs [30]. Hadoop framework has a master–slave architecture and built-in fault tolerance capability by making three or user defined replica of data nodes. Our platform provides the high performance computing over the commodity hardware of cloud infrastructure OpenStack [31]. Our platform utilized MapReduce framework that is a distributed, parallel processing architecture and uniquely qualified to exploit big data potential [32]. A MapReduce job comprises of two parts, (i) a map part, which takes raw data and organizes it into key/value pairs, and (ii) a reduce part, which processes data in parallel. This component also contains tools for data reading, writing, movement and interaction, such as Flume [29], kafka [33], Sqoop [34] and Hive [35].

9.3.4   Learning Models

The proposed learning models component is capable to learn from the large collected data and use that knowledge to predict future trends, behaviors, and decisions regarding unknown future events. Learning models contains feature bank and machine learning models. In our feature bank component, we extract the relevant feature according to the data source. Feature extraction is a highly domain-specific technique that defines a new attribute using the raw signals to reduce computational complexity and enhance the recognition process. For instance, in case of accelerometer signal we extracted the time and frequency domain features [36]. Learning model contains standard algorithms of recognizing the human contexts and behavior including non-parametric nearest neighbor model [37], evolutionary fuzzy model [38], social media processing API [39] for classification and prediction. The details about the learning models can be found in our recent publications [40].

At this point, data will be transforming to meaningful information and can provide directly to analytical service for visualizing the human contexts and behavior patterns. In our big data processing approach, machine-learning models perform batch processing in which all training data set is read once and learned parameters are stored in knowledge repositories. We stored parameters in knowledge bases instead of HDFS because frequent I/O operations can become very expensive in terms of time.

9.3.5   Model Interface

We are introducing the model interface that represents a standard interface to provide the linkage between components. It can communicate with knowledge bases as well as communication between big data storage and processing component for getting the training data. When new data arrive to the platform, or users want to retrain the learning model parameters, it will be coordinated with model interface. We can move learning models information and knowledge repositories into HDFS if the size increases from gigabytes to petabytes. This component is also responsible to maintain and modify the learning parameters when new data dimensions arrive.

9.3.6   Knowledge Bases

Once the data is distilled and processed through learning models it is loaded into information repositories, so users have cost effective and real time access to it. It also contains the information, knowledge, and Meta repositories. Knowledge repositories are filled after the inferencing module of our platform. Knowledge repositories assist the analytical services for visualization as well as reasoner and inferencing services for better quality of decisions as well as reasoning about the certain situations. In case of Meta repositories, it contains the metadata that contains the schema information and information about the integrity constrains. It will be helpful for the other APIs' consumer engineers.

9.3.7   Reasoner and Inferencing Services

This component assists the platform to provide personalized recommendations and reasoning for provided guidelines. Traditionally, reason and inference module focus on general recommendations applicable to a community of users, but not specific to each individual and their personal preferences. To accommodate personalization concept and dynamic user's query support at run time, a hybrid reasoning architecture is proposed, exploiting different approaches, such as rule-based reasoning [41], preference-based reasoning [42] and probabilistic inferencing [43]. Initially, rules are extracted explicitly from the domain knowledge and coarse grain by user preferences. To make final decision probabilistic, Bayesian network is utilized to gain certain confidence. The system can be tuned by the preference of the user collected at the initial configuration time that whether to use a single or a combination of the approaches or a specific reasoning method to generate the prompt response to their queries.

9.3.8 Analytical Services

Analytical services provide the visualization and new insight of data to uncover hidden patterns, unknown correlations and users' behavior. In certain situations, experts need current information while in other scenarios they desire historical information along with the current information. We placed Hive-based queries looking inside the data and analytical services by providing the web interfaces. The tools include d3 [44], ggplot [45], matplotlib [46], and Google charting [47], to mention just some. It can support the experts to prepare the better and effective recommendation plan for the users.

9.3.9 u-Lifecare Services API

The decision support system can be accessed via a user interface and service API to build the requested applications over the constructed knowledge base. The objective of personalized u-lifecare services is to provide timely and accurate services to the individuals based on the constructed knowledge, user's generated data as well as historical data. This represents the top layer in Fig. 9.1, and linked directly to the analytical services, and reasoner and inferencing services.

9.4 Case Study

Consider the monitoring and tracking of user's behaviour routine of a focused group or nationwide. User's behaviour can be divided into active or sedentary. While sedentary behaviour is increasing due to societal changes and related to prolonged periods of sitting. Sitting while watching television, using the computer while working or playing games for long hours, are examples of sedentary behaviours that are currently common worldwide [48]. These kinds of activities increase sedentary behaviour across all age groups. A person is considered sedentary if they spend large amount of their day with such activities and do not spend sufficient time for physical activity or exercise. Similarly, many jobs require people to sit in front of computer all the day, which also promotes sedentary behaviour. Sedentary behaviour is associated with poor health outcomes, including the high risk of overweight and obesity [49], physiological and psychological problems [50], heart disease and diabetes [51]. To promote healthy behaviour, there should be some efficient mechanisms to track and estimate the time spent in active and sedentary activities. In order to track user's behaviour in daily routines, we developed fundamental contexts tracking application based on embedded sensor of smartphone [38, 52]. Our application is capable to run in the background while users can use their smartphone for other tasks. We construct the feature bank by extracting the relevant features according to the type of sensor. During the classification phase, stored features from the codebook are loaded into the learning model and classify the current situation. Figure 9.2 shows example scenes of fundamental contexts of human behaviour.

For discussion, consider the weight management scenario for people, like Ms. Aliza. She is a 28-years-old lady, who wants to adopt an active lifestyle in her daily routines. She preferred physical exercise, such as brisk walking and jogging. Her recent weight gain has prompted her to adopt physical activities in daily routines, which can be squeezed into daily schedule with ease. She needs guidelines and recommendations that fit in her busy schedule. Statistics about her body mass index (BMI) is summarized in Table 9.3.

Aliza is interested to know about her lifestyle primitive statistics, which tells her how much physical activity or sedentary behaviour she did in the previous days as well as recommendations and routine plan in terms of daily routines to achieve her active lifestyle goal. Supporting this scenario, the proposed platform can help her in a truly ubiquitous manner to log her daily routines. She installed our developed application and creates her profile. She can maintain her personal profile; change the preferences according to the seasonal changes and system can generate personalized recommendation according to new preferences. Our application can run in the background and log the routines while using the smartphone normally. She can share her workouts over the social networks with family and friends. The sharing feature of our platform is optional and has great potential to motivate Aliza in terms of appreciation from the social networks while promoting an active

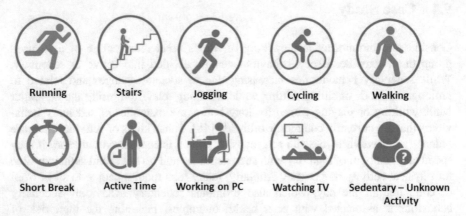

Fig. 9.2 Example of active behavior 'Stairs', 'Running', 'Jogging', 'Cycling', 'Walking' and example of sedentary behavior 'short break', 'working on PC', 'watching TV', and 'Sedentary—Unknown Activity'

Table 9.3 User's physical statistics

Gender	Age (Years)	Height (cm)	Weight (lbs)
Female	28	162	160

Table 9.4 Twenty four hour routine with time spent duration

Activity	Duration (h)
Active	3.75
Walking	0.30
Stairs	0.05
Cycling	0.15
Short breaks	0.50
Working on PC	3.58
Watching TV	1.55
Sedentary—context unknown	14.12

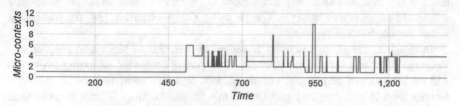

Fig. 9.3 Daly routine with time spent of activity pattern

lifestyle. Consequently, it motivates the other individuals to adopt healthy routines to keep themselves healthier. We are presenting her 24 h data of a working day. Table 9.4 shows one-day activities along with micro-contexts.

In Table 9.4, we showed the quantification of the amount of time spent in sedentary behavior by subject that was around 19 h and 15 min. In the contexts class label "Sedentary—context unknown", we consider the subject sleeping time and all other micro-contexts such as subject went to library for studying or any other that is not included in the recognized micro-context. Even though short break time is 30 min but important to indicate the state of being active. In order to get insight of the sedentary patterns of daily routines, we showed observed pattern from 12:00 to 23:59 by presenting each minute of 24 h in Fig. 9.3, where x-axis shows the time in minute while y-axis shows the micro-context and annotation as shown in Table 9.5.

Table 9.5 Twenty four hour routine with time spent duration

Activity	Class label
Sedentary	1
Working on PC	2
Watching television	3
Active	4
Short break	5
Walking	6
Running	7
Stairs	8
Jogging	9
Cycling	10

This micro-context of sedentary behaviour provide better understanding of users' daily routines and may help users to minimize the amount of prolonged sitting and adopt active lifestyle. We are using Google charting [34] for presenting behaviour analytics.

In Fig. 9.3, the subject activity is sedentary because of night and subject was sleeping. After that we can see active patterns, using computer, watching television and on off state of active and sedentary. We classify "short breaks" if subject's activity time is less than or equal to one minute and "active" if time is more than one minute. For instance, we also show the activities along starting time and sequence in Fig. 9.4.

In Figs. 9.5 and 9.6, daily routine pattern is visualized and provide us information on percentage of time in 24 h spent in different micro-contexts activities. We also extract the information about the number of short breaks (i.e., 15 short breaks) that help subject to avoid longer sedentary activity.

Our platform also provides recommendations with the help of reasoner and inference service module. It correlates her personal profile along preferences with daily routines and suggest routine plan to adopt healthy behaviour. For instance, if reasoner assesses Ms. Aliza in prolonged sitting state while watching television. Our platform generates the appropriate physical activity to complete the routine plan of the users. The inference service recommends her the following recommendations over the smartphone as a toast message with beep.

> To be active, you can take a brisk walk of 15 min. Brisk walking helps you to reduce body fat, depression, and give you stronger bones.

She can also query about the weather conditions or any further recommendations that she require to go out for recommended activity. The execution process of our platform is illustrated in the sequence diagrams of Fig. 9.7.

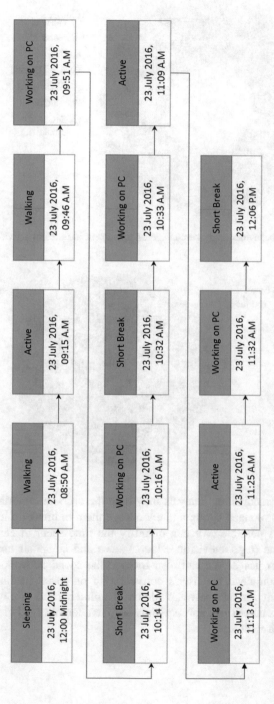

Fig. 9.4 Context recognition along with activity starting time and sequence of occurrences for 12 h

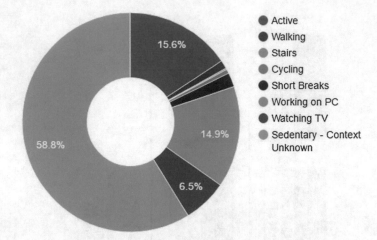

Fig. 9.5 Sedentary behaviour along percentage of time spent in different activities

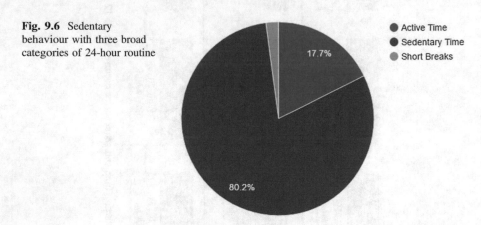

Fig. 9.6 Sedentary behaviour with three broad categories of 24-hour routine

Our behaviour recognition analysis may help the users to minimise the amount of time spent in prolonged sitting and encourage them to break up long periods of sitting as often as possible. We can quantify the time spent in electronic media during leisure time (e.g., television, video games and computer use) and set the limit hour of their usage. We can also include the social networks to know the activities and recommend the user's active group on social media, to get more knowledge, and educate own selves while overcoming barriers of physical distance or geographic isolation. It will consequently help to adopt the healthier lifestyle and reduced the health risks.

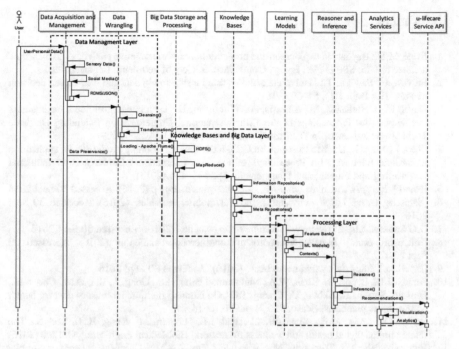

Fig. 9.7 Sequence diagram of execution process in case study

9.5 Conclusion

We proposed a platform that is able to process the structured and unstructured data gathered from multiple sources through big data technology, to provide consolidated services and analytics to assist in decision making. Our proposed platform includes number of modules and sub-modules that are capable of data collection and management, extract the knowledge and information while exploiting big storage technology for massive storage of data. For continuous sensing and acquiring of personal data, we utilized ubiquitous nature of cloud computing infrastructure. Furthermore, it provides high performance computing for intensive data processing in cost effective manner. The proposed platform proved systematic data management and effective utilization of the users' generated data that can help the individuals to visualize the personal behaviour patterns and provided *u*-lifecare services to manage their daily routines and remain active. Our future plan includes providing a comparative analysis for individuals based on certain parameters while preserving the privacy and security of the data.

References

1. Brule, M.R.: Big data in exploration and production: real-time adaptive analytics and data-flow architecture. In: SPE Digital Energy Conference. Society of Petroleum Engineers (2013)
2. Provost, F., Fawcett, T.: Data science and its relationship to big data and data-driven decision making. Big Data 1(1), 51–59 (2013)
3. Dayal, U., Castellanos, M., Simitsis, A., Wilkinson, K.: Data integration flows for business intelligence. In: Proceedings of the 12th International Conference on Extending Database Technology: Advances in Database Technology, pp. 1–11. ACM (2009)
4. Shen, J., Xing, J., Xu, M.: Research on CBR-RBR fusion reasoning model and its application in medical treatment. In: Proceedings of the 21st International Conference on Industrial Engineering and Engineering Management, pp. 431–434 (2015)
5. [Nike+]. https://www.nike.com/us/en_us/c/nike-plus/nike-app (2016). Accessed 30 Sept 2016
6. [Samsung Gear]. http://www.samsung.com/us/mobile/wearables/ (2016). Accessed 30 Sept 2016
7. [LG Smartwatch]. http://www.lg.com/us/smart-watches (2016). Accessed 30 Sept 2016
8. [Microsoft Band]. https://www.microsoft.com/microsoft-band/en-us (2016). Accessed 30 Sept 2016
9. [Fit Blaze]. https://www.fitbit.com/blaze (2016). Accessed 30 Sept 2016
10. Amin, M.B., Banos, O., Khan, W.A., Muhammad Bilal, H.S., Gong, J., Bui, D.M., Cho, S.H., Hussain, S., Ali, T., Akhtar, U., Chung, T.C.: On curating multimodal sensory data for health and wellness platforms. Sensors 16(7) p. 980 (2016)
11. Banos, O., Amin, M.B., Khan, W.A., Afzal, M., Hussain, M., Kang, B.H., Lee, S.: The mining minds digital health and wellness framework. BioMedical Engineering OnLine (2016)
12. Rawassizadeh, R., Tomitsch, M., Wac, K., Tjoa, A.M.: UbiqLog: a generic mobile phone-based life-log framework. Pers. Ubiquit. Comput. 17(4), 621–637 (2013)
13. Kambatla, K., Kollias, G., Kumar, V., Grama, A.: Trends in big data analytics. J. Parallel Distrib. Compu. 74(7), 2561–2573 (2014)
14. Rodríguez-Mazahua, L., Rodríguez-Enríquez, C. A., Sánchez-Cervantes, J. L., Cervantes, J., García-Alcaraz, J. L., Alor-Hernández, G. A.: General perspective of Big Data: applications, tools, challenges and trends. J. Supercomputing pp. 1–41 (2015)
15. Rodríguez-García, M.Á., Valencia-García, R., García-Sánchez, F., Samper-Zapater, J.J.: Creating a semantically-enhanced cloud services environment through ontology evolution. Future Gener. Comput. Syst. 32, 295–306 (2014)
16. Ling, T., Kang, B. H., Johns, D. P., Walls, J., Bindoff, I.: Expert-driven knowledge discovery. In: IEEE Fifth International Conference on Information Technology: New Generations pp. 174–178 (2008)
17. Rospocher, M., Serafini, L.: An ontological framework for decision support. In: Joint International Semantic Technology Conference pp. 239–254 (2012)
18. [IBM].   https://www-01.ibm.com/software/data/bigdata/what-is-big-data.html   (2016). Accessed 8 July 2016
19. Gupta, P., Dallas, T.: Feature selection and activity recognition system using a single triaxial accelerometer. IEEE Trans. Biomed. Eng. 61(6), 1780–1786 (2014)
20. Hall, M., Frank, E., Holmes, G., Pfahringer, B., Reutemann, P., Witten, I.H.: The WEKA data mining software: an update. ACM SIGKDD Explor. Newslett. 11(1), 10–18 (2009)
21. Theekakul, P., Thiemjarus, S., Nantajeewarawat, E., Supnithi, T., Hirota, K.: A rule-based approach to activity recognition. In: Knowledge, Information, and Creativity Support Systems, pp. 204–215. Springer, Berlin (2011)
22. Furche, T., Gottlob, G., Libkin, L., Orsi, G., Paton, N.W.: Data wrangling for big data: challenges and opportunities. In: 19th International Conference on Extending Database Technology (EDBT). Bordeaux, France (2016)
23. Rahm, E., Do, H.H.: Data cleaning: problems and current approaches. IEEE Data Eng. Bull. 23(4), 3–13 (2000)

24. Michalski, R.S., Carbonell, J.G., Mitchell, T.M. (eds.): Machine Learning: An Artificial Intelligence Approach. Springer Science & Business Media (2013)
25. [Data Wrangler]. http://vis.stanford.edu/wrangler/ (2016). Accessed 28 July 2016
26. [Tabula]. http://tabula.technology/ (2016). Accessed 28 July 2016
27. [Openrefine]. http://openrefine.org/ (2016). Accessed 28 July 2016
28. [Python and Pandas]. http://pandas.pydata.org/ (2016). Accessed 28 July 2016
29. [Apache Flume]. http://flume.apache.org/ (2016). Accessed 28 July 2016
30. [Apache Hadoop]. http://hadoop.apache.org (2016). Accessed 28 July 2016
31. Sefraoui, O., Aissaoui, M., Eleuldj, M.: OpenStack: toward an open-source solution for cloud computing. Int. J. Comput. Appl. **55**(3) (2012)
32. Dean, J., Ghemawat, S.: MapReduce: simplified data processing on large clusters. Commun. ACM **51**(1), 107–113 (2008)
33. [Apache Kafka]. http://kafka.apache.org/ (2016). Accessed 28 July 2016
34. [Apache Sqoop]. http://sqoop.apache.org/ (2016). Accessed 28 July 2016
35. [Apache Hive]. http://hive.apache.org/ (2016). Accessed 28 July 2016
36. Fahim, M., Lee, S., Yoon, Y.: SUPAR: Smartphone as a ubiquitous physical activity recognizer for u-healthcare services. In: 2014 36th Annual International Conference of the IEEE Engineering in Medicine and Biology Society, pp. 3666–3669 (2014)
37. Fahim, M., Khattak, A.M., Chow, F., Shah, B.: Tracking the sedentary lifestyle using smartphone: a pilot study. In: 2016 18th International Conference on Advanced Communication Technology (ICACT), pp. 296–299 (2016)
38. Fahim, M., Fatima, I., Lee, S., Park, Y.T.: EFM: evolutionary fuzzy model for dynamic activities recognition using a smartphone accelerometer. Appl. Intell. **39**(3), 475–488 (2013)
39. [Alchemy API]. http://www.alchemyapi.com/social-media-monitoring (2016). Accessed 28 July 2016
40. Fahim, M., Idris, M., Ali, R., Nugent, C., Kang, B., Huh, E.N., Lee, S.: ATHENA: a personalized platform to promote an active lifestyle and wellbeing based on physical, mental and social health primitives. Sensors **14**(5), 9313–9329 (2014)
41. Sun, R.: Robust reasoning: integrating rule-based and similarity-based reasoning. Artif. Intell. **75**(2), 241–295 (1995)
42. Ali, R., Afzal, M., Hussain, M., Ali, M., Siddiqi, M.H., Lee, S., Kang, B.H.: Multimodal hybrid reasoning methodology for personalized wellbeing services. Comput. Biol. Med. **69**, 10–28 (2016)
43. Murphy, K.P.: Dynamic bayesian networks: Representation, inference and learning (Doctoral dissertation, University of California, Berkeley) (2002)
44. [D3]. https://d3js.org/ (2016). Accessed 28 July 2016
45. [ggplot2]. http://ggplot2.org/ (2016). Accessed 28 July 2016
46. [matplotlib]. http://matplotlib.org/ (2016). Accessed 28 July 2016
47. [Google charting]. https://developers.google.com/chart/ (2016). Accessed 28 July 2016
48. Dunstan, D.W., Healy, G.N., Sugiyama, T., Owen, N.: Too much sitting: the population health science of sedentary behavior. Eur. Endocrinol. **6**(1), 19–23 (2010)
49. Tremblay, M.S., Colley, R.C., Saunders, T.J., Healy, G.N., Owen, N.: Physiological and health implications of a sedentary lifestyle. Appl. Physiol. Nutr. Metab. **35**(6), 725–740 (2010)
50. Barwais, F.A., Cuddihy, T.F.: Empowering sedentary adults to reduce sedentary behavior and increase physical activity levels and energy expenditure: a pilot study. Int. J. environ. Res. Public Health **12**(1), 414–427 (2015)
51. Vandelanotte, C., Duncan, M.J., Short, C., Rockloff, M., Ronan, K., Happell, B., Di Milia, L.: Associations between occupational indicators and total, work-based and leisure-time sitting: a cross-sectional study. BMC Public Health **13**(1), 1 (2013)
52. Fahim, M., Khattak, A.M., Thar, B., Chow, F., Shah, B.: Micro-context recognition of sedentary behaviour using smartphone. In: 2016 6th International Conference on International Conference on Digital Information & Communication Technology & its Applications (DICTAP2016) (2016)

Chapter 10
Decision Support System for Operational Risk Management in Supply Chain with 3PL Providers

Juan Carlos Osorio Gómez, Diego Fernando Manotas Duque, Leonardo Rivera and Jorge Luis García-Alcaraz

Abstract For organizations, it is essential to make good decisions regarding risks present both in the company and in its supply chain. The risk management process includes risk identification, evaluation, prioritization and quantification. The participation of Third Party Logistics providers (3PL) in supply chains has been increasing, and it is important to consider how their presence affects risk management. Finally, a decision support system helps decision makers to process, analyze and define actions related to large amounts of data. We present a multi-criteria decision support system for effective management of the operational risks present in a supply chain that includes 3PL providers, specifically in ground transportation of goods. The model is supported by Fuzzy QFD for the prioritization of risks in terms of their impact on the performance indicators that are considered relevant by the actors in the supply chain. Findings indicate that the proposed model allows to prioritize the risks according with the most important indicators and for the case of study we found the most critical risks.

Keywords Decision support systems · Operational risk management · Fuzzy QFD · Supply chain risk management · Third party logistics - 3PL

10.1 Introduction

Supply chain management is currently one of the most important subjects in practice and academia, especially because of the change from local vision to a global outlook of markets, and also due to the integrated point of view that changes

J.C. Osorio Gómez (✉) · D.F. Manotas Duque · L. Rivera
Grupo de Investigación en Logística y Producción,
Escuela de Ingeniería Industrial, Universidad del Valle, Cali, Colombia
e-mail: juan.osorio@correounivalle.edu.co

J.L. García-Alcaraz
Departamento de Ingeniería Industrial y Manufactura, Instituto de Ingeniería
y Tecnología, Universidad Autónoma de Ciudad Juárez, Ciudad Juarez, Mexico

© Springer International Publishing AG 2017
G. Alor-Hernández and R. Valencia-García (eds.), *Current Trends on Knowledge-Based Systems*, Intelligent Systems Reference Library 120, DOI 10.1007/978-3-319-51905-0_10

the paradigm from the study of individual nodes on a network to a more global vision. One of the specific areas under scrutiny relates to the management of risks in the chain. This has importance for both academics and practitioners alike. The need to have a risk management system that enables companies and networks to identify, evaluate, quantify and manage risks is no longer under discussion. Companies pay more attention to risks and improve their information systems to include facts related to them. This accentuates the need to handle larger volumes of information related to risks in order to manage them more effectively.

In this context, decision support systems become more important, because they give decision makers criteria to make decisions more effectively, that is, to exploit more productively the resources available, to make decisions more quickly, and to have the results of those decisions bring a positive impact to the organization.

The increasing need of the companies to focus on the core business object has generated a trend oriented to outsource different activities. In this context, supply chain activities have evolved from a first stage where we have companies that are responsible for their logistics processes up to the current trend with companies who have delegated all their logistics activities to specialized agents [1]. We will consider as 3PLs those organizations devoted to offer fundamental logistics services to companies, in a way that enables them to focus on their core businesses.

This chapter presents our proposal for a decision support system for the management of operational risks in supply chains with 3PL providers.

10.2 State of the Art

A Decision Support System can assist a decision maker in processing, assessing, categorizing and organizing information in a useful fashion that can be easily retrieved in different forms. In other words, a DSS is a computer technology solution that can be used to support complex decision making and problem solving [2].

According with [2] the original DSS concept was most clearly defined by Gorry and Morton in 1971 who combined categories of management activities developed by Anthony in 1965 with description of decision types proposed by Simon in 1960 using the terms structured, semi-structured and unstructured rather than programmed and non- programmed.

We did not find works related with decision support systems for operational risk management in supply chain that involves 3PL providers in the literature. We think such a system is required because 3PL providers are increasing their participation in supply chains and they are directly related with risk in supply chain.

There are many papers related with decision support systems and risk management such as [3], who present a quantitative approach to construction risk management through AHP and decision tree analysis. Also, [2] developed a risk management DSS for international construction projects, [4] proposed a fuzzy decision support system (FDSS) for the assessment of risk in e-commerce development. [5] presented a DSS that can provide information on the environmental impact of anthropic activities by examining their effects on groundwater quality using fuzzy logic. [6] proposed a DSS for risk analysis on dynamic alliance. [7] presented a unified framework for developing a DSS or supporting the management of emergency response operations, [8] presented a DSS for the modeling and management of project risks and risk interactions. [9, 10] present operational risk management like new paradigm for decision making and [11] present a conceptual knowledge approach to operational risk management.

Finally, there are some important DSS applications for the supply chain management like; as for example, [12] proposed a middleware-oriented integrated architecture with the use of semantic features as data provider to offer a brokerage service for the procurement of products in a SCM, [13] proposed a new software architecture named SKOSCM to offer a brokerage service for e-procurement in supply chains and [14] present a system dynamics model to evaluate scenarios that improve the performance of the automotive supply chain.

10.2.1   Supply Chain Risk Management—SCRM

According to [15] risk in the supply chain is the potential loss of a Supply Chain (SC) in terms of its objective values of efficiency and efficacy due to the uncertain evolution of the characteristics of the SC, whose changes were caused by trigger events.

Due to the influence on risk in the logistical performance, the implementation of risk management has become a critical aspect. This management can be viewed from the point of view of mitigation or contingency [16].

Effective risk management has become a focal issue in organizations in order to survive in their competitive environments. Thus, SCRM has emerged as a natural extension of Supply Chain Management (SCM) with the primary objective of identifying potential sources of and developing action plans for their mitigation [17].

An effective system for supply chain risk management has to identify, evaluate and quantify risks in such a way that the organization is able to generate its plans according to the risks that have bigger effects on their corporate objectives.

In this sense, and according to the literature review, we present our approach for supply chain risk management in chains with 3PL in Fig. 10.1.

Fig. 10.1 Supply Chain Risk Management System [1]

10.2.2 Operational Risk

Risk is a measure of random fluctuations in performance through time. Operational risk measures the connection between those performance fluctuations and business activities [18].

Operational risk is as old as businesses. Each industry, each activity faces operational risks [19].

In summary, operational risks include most of what can harm the organization. Operational risks are foreseeable, and to some measure, avoidable (if not the adverse event, at least its consequences on the organization). It is clear that operational risks might be mitigated only after they have been correctly identified. A risk that has been correctly identified is no longer a risk, it becomes a management problem [20].

Our proposal includes mainly operational risks related to 3PL activities that affect the supply chain. Table 10.1 shows some related papers where operational risk is discussed, according to the literature review.

10.2.3 Outsourcing as a Logistics Strategy

Outsourcing is being used as an important managerial strategy in the world. It enables companies to focus in their core activities and leave the non-core activities to specialized third-parties. Some of the advantages of outsourcing include cost reduction, lead time reduction, decreasing the time required for product development, improvements in customer service, and the reduction of total logistics costs [40–42].

Current supply chains have evolved beyond the traditional configuration of supplier—manufacturer—consumer (Fig. 10.2) to a chain where 3PL providers are integral links (Fig. 10.3).

To operate successful supply chains it is necessary to have successful logistics activities. This is why the role of 3PL providers has switched from a few simple tasks to a total outsourcing, evolving from a simple provider of logistical services to

Table 10.1 Papers where operational risk is discussed

Risk	Authors
Operational risk	[21–39]

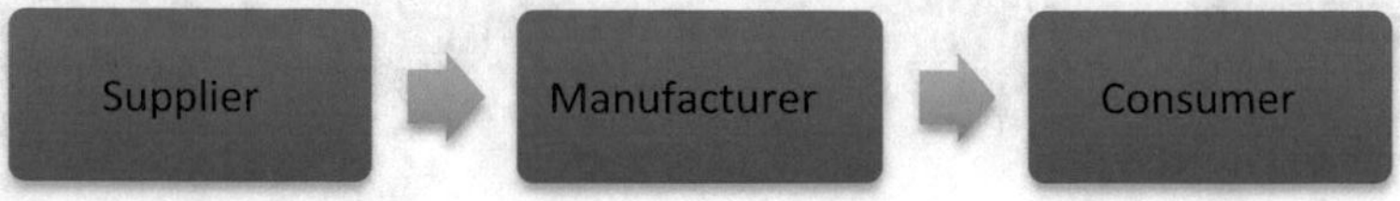

Fig. 10.2 Basic configuration of the supply chain

Fig. 10.3 Supply chain with the participation of the 3PL providers (adapted from [43])

a strategic provider in the supply chain, working simultaneously with several partners in it [44].

The role of logistics services is critical, because it drives the flow of materials and information in the supply chain, forward and backwards. The importance of logistics outsourcing in global supply chains becomes more notorious with the growth of companies and their worldwide expansion [44].

The growing need of companies to focus on the core object of their business has originated a trend towards outsourcing different activities. In this context, the activities of the supply chain have evolved from a first stage with companies responsible for their own logistical processes, to the current trend of companies delegating all their logistical activities to specialized agents.

According to [45], the most important services offered by 3PL providers and the percentage in which they are used are: Domestic transportation 80%, Storage 66%, International transportation 60%, Cargo transportation 48%, Customs brokerage 45%, Reverse logistics 34% and Cross-docking 33%.

10.2.4 Risk in 3PL Operations

According to the Council of Supply Chain Management (CSCM), logistics management must deliver "The Right Product, To the Right Customer, At the Right Time, At the Right Place, In the Right Condition, In the Right Quantity and Quality, At the Right Cost". It is not easy to coordinate all of these goals for all the internal and external partners. Therefore, to be able to deliver in all these dimensions, a global enterprise needs to support itself in the collaboration of its partners, both upstream and downstream [16].

Even though studies show great benefits for companies that outsource their logistics activities through 3PL providers, it is also true that this outsourcing increases risks in the supply chain. Therefore, risks in outsourcing activities need to be carefully considered.

Lee et al. [46] argue that besides recognizing the advantages of outsourcing, companies should also consider the risk of making bad decisions regarding outsourcing. They contend that an enterprise can externalize different functions, and this implies a different degree of commitment and integration between the company and its outsourcing service suppliers. When the degree of commitment varies, the complexities of the interaction will surface, creating different issues of risks.

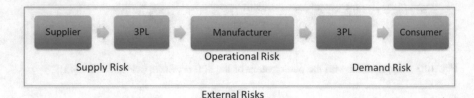

Fig. 10.4 Risks in a supply chain with 3PL providers (*Source* [43])

Therefore, any outsourcing decisions must fully consider the potential risks that go along with them.

Finally, to configure the risks in a supply chain with 3PL providers, Fig. 10.4 presents the risks according to their source ([43]). This classification includes risks that are internal to the company, external to the company but in the supply chain and risks that come from outside the supply chain. In this proposal we have considered the first two classes of risks, which, according to [47], are the ones that require more attention in the relationships of outsourcing.

10.3 Proposed Model

We propose a decision support system for managing operational risks that considers the phases presented in Fig. 10.1 and is configured according to Fig. 10.5. Each of its phases will be presented in detail and accompanied by an example of application in the following section.

The idea is to have a database of all the possible risks associated with each of the activities under consideration. Then, we will establish priorities applying the multi-criteria tool Fuzzy QFD, according to the impact that each of the risks may have on the strategic indicators of the company. Once the priorities are established, we will quantify the risks using the appropriate technique according to information availability and we will finally outline managerial plans that eliminate or mitigate the risks under consideration.

This process should have a cyclical nature, in such a way that the organization performs a periodic evaluation and updates its risks, as well as the measures to face them. The system should also extend to the whole supply chain, to share information between the organizations involved in such a way that more information is available for study and analysis. In this sense, it is possible that 3PL providers do business with more than one organization in the supply chain, thus making this sharing of information advantageous for the unification of strategies between these providers and the other links in the supply chain.

The information to share will be related to the risks identified and their associated costs. If it is necessary to consider the prioritization and the action plans as confidential material, the system will ensure access control. The system will enable

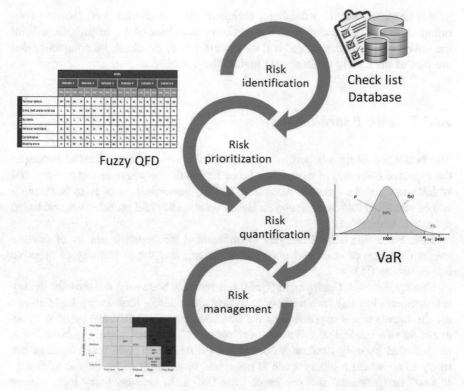

Fig. 10.5 Decision support system for operational risk management in the supply chain

the sharing of the information related to the 3PL provider regardless of which of the other links in the supply chain has a relationship with it, thus strengthening the database.

10.3.1   Risk Identification

The first step in risk management is the identification of the sources of risk. Companies need to detect every possible threat systematically. Risks can reside in different parts of the operation, both internal and external. In supply chains, the possibility of risk exposure is higher than in other activities [16]. The identification of risks is an important first step in any risk management effort [17].

There is a variety of techniques that help in the identification of risks. Among these we can count supply chain mapping, checklists, event tree analysis, fault tree analysis, failure modes and effects analysis (FMEA) and Ishikawa's cause and effect diagram [48]. People with the right knowledge and expertise should be involved in risk identification [49].

We propose to build a database, enriched with the literature review, incorporating existent risks according to the particular activities of the organization. With the arrival of new organizations and the occurrence of events, those companies that are part of the supply chain should update the database.

10.3.2   Risk Prioritization

The evaluation of risks is performed to define the actions to be taken according to the expected outcomes of these risks. It is a tool to define where to focus efforts and which risks may be ignored. According to [50] quantification of risks in research can be classified into two categories: Based on statistics and probabilities and based on expert knowledge.

Risk evaluation helps managers to understand the negative impact of adverse events (in terms of costs and poor performance), and the probability of negative consequences ([51]).

Quality Function Deployment (QFD) has recently been used not only for quality management, but also as a multi-criteria decision making tool. Fuzzy logic allows for the mathematical expression of the intermediate values that can be used by an evaluator of a qualitative "situation" or "problem", in cases where the evaluator is not satisfied by only Boolean values (false [0] or true [1]). In real life, there are many cases where a binary scale is not enough, rather it has a degree of "truthfulness" or "falseness" that can range from 0 to 1. In essence, fuzzy logic widens the options when facing a situation, because it open an interval between 0 and 1 to emit judgment.

The methodology we will present is fundamentally supported on the work of [52]. Some other authors have used fuzzy QFD for risk management: [53, 54].

We will define strategic indicators associated to the process under evaluation. These indicators will be weighted using fuzzy logic according to the knowledge and expertise of the decision-making team.

Once the weights of the indicators are defined, they will be evaluated and their impact will be assessed by the same team (using the information system), in such a way that the QFD methodology determines a priority value for each of them.

10.3.3   Risk Quantification

Bocker y Kluppelberg in 2005 argue that the only feasible way to successfully manage operational risks is to identify and minimize them, which requires the development of the right techniques for their quantification [55].

In order to evaluate and compare different solutions that limit the impact of risks, decision makers have to (somehow) quantify the risk. Standard deviation, half-variance approaches, value at risk, conditional value at risk or risk premiums are measures that have the objective to describe the interaction between uncertainty and the extent of their related damage or benefit. Due to the lack of quantitative measurements that capture the most complex realities of supply chains, these measures (developed in finance and insurance contexts) are also applied for supply chain risks [15].

The information system will support the quantitative data related to the economic effects of risks in such a way that a quantification of them becomes possible. With the evolution of the system and the integration of links in the supply chain to it, more information will be available and better analyses will become feasible.

10.3.4   Risk Management

Due to the influence of risk in logistics performance, implementing risk management systems has become a critical aspect. This management can be understood from the point of view of mitigation or the contingency approach [16].

It is then necessary to establish the actions required to eliminate the risk or mitigate its consequences, according to the quantification performed. In this sense, it is important to establish an action plan that clearly defines responsible parties and to perform periodic reviews, supported with the information system that has been established. Once a risk has been eliminated or mitigated, it is necessary to update the database and to continue the process with the risks that follow in the list of descending priority, in order to minimize their impacts.

The model suggests a continuous process where the four phases are conducted periodically in such a way that the database of existing risks is updated, existing prioritizations are validated, action plans are followed upon and companies have up-to-date and online information about risks and action plans.

As an example, we present a case study where the proposed system has been applied to a company in Colombia. This company is part of a supply chain with 3PL providers, and the system has been used for risks associated to domestic ground transportation of products.

10.4   Case Study

The case study is related to a supply chain that works with food products, especially baked goods. It is specifically related to the manufacturing company, which requires a raw material that needs to be transported in special conditions, because its temperature is a critical factor in the success of the manufacturing process. Transportation of this material is done by 3PL providers. We will show the decision support model associated to the risks that are present in the transportation activities of this raw material.

10.4.1 Risk Identification

For the initial identification, we performed a literature review related to the most common risks related to the activity of interest. We also conducted interviews with people in the organization who are in charge of the organization and the supply chain. These risks are entered in the database in such a way that they can be accessed through the information system. The people in charge of risk management select from the database the risks related to the activity under evaluation. Table 10.2 shows the risk from the database (they are obtained from literature review and most of them were obtained from the company) and Fig. 10.6 presents the risk selection process with the risks that were identified for the case study highlighted in bold. These are the risks that decision makers of the company defined from the database. In the event that new risks are identified, they must be entered into the database. This update can be performed by any echelon of the supply chain, according with the architecture of the DSS (Fig. 10.7).

10.4.2 Risk Prioritization

We used Fuzzy QFD for the prioritization and evaluated the impact of the identified risks on the strategic objectives of the organization. We use the fuzzy scale presented in Table 10.3 for the evaluation, which has triangular fuzzy numbers to represent the linguistic expressions employed by the people in charge of the evaluation process. These experts evaluate the impact the risks identified in the previous step have in the performance indicators associated with the process under study, in this case, the ground transportation of products. Table 10.5 shows the

Table 10.2 Risk in transportation activities

Risks	
Lack of experience	Accidents
Absence or bad communication between driver and owner	Contamination
Incorrect documentation	Cross-contamination in transportation
Lack of procedures	Shipping errors
Technical defects	Driver lack of skills
Disruptions in the cold chain	Non-compliance with traffic laws
Bad road conditions	Vehicle breakdowns
Crime, theft and terrorist acts	Strikes, public demonstrations, riots
Delays due to labor strikes	Road-affecting disasters
Delays due to different handling of police inspections	Goods damaged
Vehicular restrictions	Drivers' diseases

Risks	Selection
Technical defects	⊙
Accidents	⊙
Lack of procedures	☐
Theft - security	⊙
Incorrect documentation	☐
Vehicular restrictions	⊙
Contamination	⊙
Lack of experience	☐
Absence or bad communication between driver and owner	☐

Fig. 10.6 Database for selecting risks (risk identification phase)

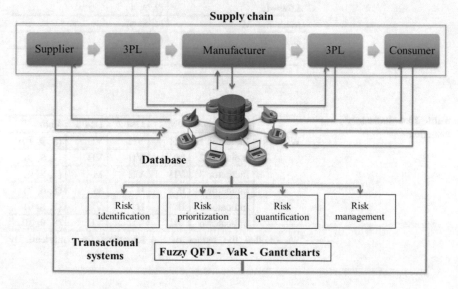

Fig. 10.7 Decision Support System Architecture

indicators defined for this case and the evaluation of their importance established by the three members of the risk management team. This evaluation is performed because not all risks are equally important for the company. This evaluation is analogous to the prioritization of the "Whats" in QFD and the scale in Table 10.3 is used for this. The results presented in Table 10.4 are obtained using fuzzy math.

Once risks, indicators and their corresponding weights have been defined, we proceed (according to the QFD methodology) to evaluate the impact that each of the identified risks has on the selected indicator, to find out which are the priority risks on which the company should focus its elimination and mitigation efforts. Figure 10.8 presents a part of the QFD analysis where we evaluate the impact the risks have on the selected indicators. This impact is evaluated by the risk management team of the company, using the linguistic scale from Table 10.4. For example the decision makers agree that the impact of technical defects in indicator 1 is very high (VH).

The final result of the application of this methodology is presented in Table 10.5, where it is clear that the risk with the highest priority for the organization is shipping errors, with technical defects in the vehicles ranked in second place. The company should focus its initial efforts on these two risks. However, it is also necessary to quantify these risks to have a clear view of the investments that are required. We will discuss this quantification in the next section.

Table 10.3 Linguistic variables for the score

Linguistic variable	Triangular fuzzy number
Very low—VL	(0, 1, 2)
Low—L	(2, 3, 4)
Medium—M	(4, 5, 6)
High—H	(6, 7, 8)
Very high—VH	(8, 9, 10)

Table 10.4 Indicators[a] valuation

	Indicators	DM 1	DM 2	DM 3	Value
WHATs	Indicator 1	VH	VH	VH	(8, 9, 10)
	Indicator 2	M	VH	VH	(7, 8, 9)
	Indicator 3	VH	VH	M	(7, 8, 9)
	Indicator 4	H	H	M	(5, 6, 7)
	Indicator 5	H	H	VH	(7, 8, 9)
	Indicator 6	VH	VH	VH	(8, 9, 10)

[a]We omitted the names of the indicators for confidentiality reasons

	Indicator 1			Indicator 2			Indicator 3			Indicator 4			Indicator 5			Indicator 6		
HOWs / WHATs	DM1	DM2	DM3	DM1	DM2	DM3	DM1	DM2	DM3	DM1	DM2	DM3	DM1	DM2	DM3	DM1	DM2	DM3
Technical defects	VH	VH	VH	M	H	H	H	M	VH	VL	L	VL	VL	VL	VL	H	VH	VH
Crime, theft and terrorist acts	L	VL	VL	VH	VH	VH	M	H	M	L	VL	VL	VL	VL	VL	M	H	M
Accidents	M	VL	L	L	VL	VL	H	VH	VH	VL	VL	VL	M	L	VL	VH	VH	VH
Vehicular restrictions	VL	VL	L	H	M	H	M	L	L	VH	VH	VH	L	VL	L	H	H	H
Contamination	VL	VL	VL	VL	VL	L	M	L	M	L	VL	VL	H	VH	VH	M	H	H
Shipping errors	H	H	VH	H	M	L	VH	VH	VH	M	H	H	VH	H	H	VH	VH	VH

Fig. 10.8 Fuzzy QFD for Risks Prioritization

Table 10.5 Risk prioritization

Risk	Triangular fuzzy number			Crisp value	Ranking order
Technical defects	219	300	393	303	2
Crime, theft and terrorist acts	124	189	266	192	5
Accidents	133	199	277	202	3
Vehicular restrictions	128	190	264	193	4
Contamination	116	179	254	182	6
Shipping errors	309	402	507	405	1

Additionally, and as a part of the decision support system for risk management, the risk evaluations are stored in the database until action plans to eliminate and mitigate them are defined. In this way there is control in the advance of the action plans. Once the action plans have been implemented, the risks will be deactivated in their priority, but historical information will be kept for future evaluations.

10.4.3   Risk Quantification

Quantification of operational risks is one of the more complex issues in risk management, due to the difficulty in securing the information required to perform it. One of the aims of the proposed system is to consolidate a database for the organization and the supply chain, to keep the information related to the costs of the risks in such a way that risk quantification is feasible.

According to [56], the OpVaR (Operational Value at Risk) represents a percentile of the distribution of losses. It is a statistical measure, and thus requires the a priori establishment of a series of parameters: An interval or level of confidence associated to its calculation; a time unit, to which the estimation will refer; a reference currency and a hypothesis about the distribution of the distribution of the variable under analysis. They conclude that the OpVaR can be interpreted as a figure, expressed in monetary units, that presents the minimum potential loss in a one-year horizon with a statistical confidence level of 99.9%.

For the case under study, we have the information related to technical defects, and we use the OpVaR to calculate the economic value associated to this risk. Table 10.6 presents the historical values and we calculate the OpVaR to be $58.863,70.

10.4.4 Risk Management

In this section we try to establish an action plan directed to eliminate or mitigate the high-priority risks. This action plan will establish what to do, when to do it, who will be in charge of its implementation, when should the actions be in place, and a budget for their implementation. All the action plans will be entered into the database for the risk management team to follow the advance in their implementation. Table 10.7 presents a fragment of the action plan proposed to deal with the risk related to Shipping Errors.

Table 10.6 Historical data for costs related to technical defects

Month	Value	Month	Value	Month	Value
1	2.679,97	13	3.308,71	25	3.422,90
2	1.345,37	14	1.456,62	26	1.661,86
3	1.256,35	15	1.309,31	27	1.203,32
4	2.928,37	16	604,57	28	3.192,30
5	2.317,00	17	2.608,03	29	1.405,07
6	2.821,83	18	2.936,28	30	2.729,50
7	2.510,82	19	724,38	31	453,71
8	1.554,17	20	1.719,01	32	781,22
9	920,47	21	2.660,10	33	2.736,95
10	953,85	22	3.908,47	34	5.024,57
11	3.145,39	23	486,86	35	3.765,48
12	1.523,86	24	461,06	36	463,47

Table 10.7 Action plan for the shipping errors

Activity	Responsible	Duration	Observations
Implementation of the new shipping process	Warehouse manager	30 days	
Training with 3PL providers	Warehouse manager	15 days	Once the process is ready
Implementation of audits for the most important shippings	Warehouse manager	Permanent	

Each of the activities will be entered into the database with its responsible parties and execution dates so the risk management team is able to control and follow upon them. They will also be stored to have historical information and to evaluate the effectiveness of the actions implemented.

10.5 Concluding Remarks

We propose a novel approach for operational risk management in supply chains that are supported on the fuzzy QFD for prioritizing risks. This allows organizations to be clear about the risks that are directly impacting their strategic indicators.

There are not similar approaches in the literature review for operational risk management in supply chains that involve the 3PL companies and that allow all the echelon in the chain to share information related with the risks and with the actions to mitigate or eliminate them.

The participation of the 3PL providers in the logistics activities of the supply chain is growing as a whole. The model we presented directly involves these providers and seeks to take advantage of their involvement with more than one of the links in the supply chain.

We propose to share information among the members of the supply chain, taking into account that having the information required for analysis and measurement is one of the most critical parts in risk management. Building a solid database with this information will enable the construction of better analyses and improved results for organizations and the supply chain.

For future research we propose to consider the interaction between risks in different 3PL activities, such as transportation with warehousing, and inventory management for example.

Future research can consider different alternatives for risk quantification, not only VaR but also advanced methods like g-h distribution and extreme value theory.

References

1. Manotas, D.F., Osorio, J.C., Rivera, L.: Operational risk management in Third Party Logistics (3PL). Handb. Res. Manag. Strateg. Achiev. Optim. Perform. Ind. Process. (2016). doi:10.4018/978-1-5225-0130-5
2. Arikan, A.: Development of a risk management decision support system. (2005)
3. Dey, P.K.: Decision support system for risk management: a case study. Manag. Decis. **39**, 634–649 (2001). doi:10.1108/00251740110399558
4. Ngai, E.W.T., Wat, F.K.T.: Fuzzy decision support system for risk analysis in e-commerce development. Decis. Support Syst. **40**, 235–255 (2005). doi:10.1016/j.dss.2003.12.002
5. Uricchio, V.F., Giordano, R., Lopez, N.: A fuzzy knowledge-based decision support system for groundwater pollution risk evaluation. J. Environ. Manage. **73**, 189–197 (2004). doi:10.1016/j.jenvman.2004.06.011

6. Li, Y., Liao, X.: Decision support for risk analysis on dynamic alliance. Decis. Support Syst. **42**, 2043–2059 (2007). doi:10.1016/j.dss.2004.11.008

7. Zografos, K.G., Vasilakis, G.M., Giannouli, I.M.: Methodological framework for developing decision support systems (DSS) for hazardous materials emergency response operations. J. Hazard. Mater. **71**, 503–521 (2000). doi:10.1016/S0304-3894(99)00096-5

8. Fang, C., Marle, F.: A simulation-based risk network model for decision support in project risk management. Decis. Support Syst. **52**, 635–644 (2012). doi:10.1016/j.dss.2011.10.021

9. Beroggi, G.E.G., Wallace, W.A.: Operational risk management: a new paradigm for decision making. IEEE Trans. Syst. Man Cybern. **24**, 1450–1457 (1994)

10. Beroggi, G.E.G., Waliace, W.A.: Multi-expert operational risk management. IEEE Trans. Syst. Man Cybern. Part C Appl. Rev. **30**, 1–34 (2007). doi:10.1109/5326.827452

11. Jebrin, A.H., Abu-Salma, A.J.: Conceptual knowledge approach to operational risk management (a case study). Int. J. Bus. Manag. **7**, 289–302 (2012). doi:10.5539/ijbm. v7n2p289

12. Alor-Hernández, G., Sánchez-Ramírez, C., Cortes-Robles, G., et al.: BROSEMWEB: a brokerage service for e-Procurement using Semantic Web Technologies. Comput. Ind. **65**, 828–840 (2014). doi:10.1016/j.compind.2013.12.007

13. Rodríguez-Enríquez, C.A., Alor-Hernández, G., Mejia-Miranda, J., et al.: Supply chain knowledge management supported by a simple knowledge organization system. Electron. Commer. Res. Appl. **19**, 1–18 (2016). doi:10.1016/j.elerap.2016.06.004

14. Sánchez-Ramírez, C., Alor-Hernández, G., García-Alcaráz, J.L., Tlapa-Mendoza, D.A.: The use of simulation software for the improving the supply chain: the case of automotive sector. Trends Appl. Softw. Eng. (2016). doi:10.1007/978-3-319-26285-7

15. Heckmann, I., Comes, T., Nickel, S.: A critical review on supply chain risk—definition, measure and modeling. Omega **52**, 119–132 (2015). doi:10.1016/j.omega.2014.10.004

16. Wee, H.M., Blos, M.F., Yang, W.: Risk management in logistics. Handbook on Decision Making, pp. 285–305. Springer, Berlin Heidelberg (2012)

17. Singhal, P., Agarwal, G., Mittal, M.L.: Supply chain risk management: review, classification and future research directions. Int. J. Bus. Sci. Appl. Manag. **6**, 15–42 (2011)

18. King, J.: Operational Risk—Measurement and Modelling. (2001)

19. Franzetti, C.: Operational Risk Modelling and Management. Chapman Hall/CRC Finance Series (2011). doi:10.1007/s13398-014-0173-7.2

20. Kenett, R., Yossi, R.: Operational Risk Management. A practical approach to intelligent data analysis. (2011). doi:10.1007/s13398-014-0173-7.2

21. Mand, J.S., Singh, C.D., Singh, R.: Implementation of critical risk factors in supply chain management. (2013)

22. Zandhessami, H., Savoji, A.: Risk management in supply chain management. Int. J. Econ. Manag. Sci. **1**, 60–72 (2011)

23. Ouabouch, L., Amri, M.: Analysing supply chain risk factors: a probability-impact matrix applied to pharmaceutical industry. J. Logist. Manag. **2**, 35–40 (2013). doi:10.5923/j.logistics. 20130202.01

24. Squire, B., Chu, Y.: Supply Chains at Risk : A Delphi Study. (2012)

25. Chopra, S., Sodhi, M.S.: Managing risk to avoid supply-chain breakdown. MIT Sloan Manag. Rev. **46**, 53–61 (2004)

26. Manuj, I., Mentzer, J.T.: Global supply chain risk management strategies. Int. J. Phys. Distrib. Logist. Manag. **38**, 192–223 (2008). doi:10.1108/09600030810866986

27. Avelar-Sosa, L., García-Alcaraz, J.L., Castrellón-Torres, J.P.: The effects of some risk factors in the supply chains performance: a case of study. J. Appl. Res. Technol. **12**, 958–968 (2014). doi:10.1016/S1665-6423(14)70602-9

28. Mitra, S., Karathanasopoulos, A., Sermpinis, G., et al.: Operational risk: emerging markets, sectors and measurement. Eur. J. Oper. Res. **241**, 122–132 (2015). doi:10.1016/j.ejor.2014. 08.021

29. Thun, J.-H., Hoenig, D.: An empirical analysis of supply chain risk management in the German automotive industry. Int. J. Prod. Econ. **131**, 242–249 (2011). doi:10.1016/j.ijpe. 2009.10.010
30. Tang, C., Tomlin, B.: The power of flexibility for mitigating supply chain risks. Int. J. Prod. Econ. **116**, 12–27 (2008). doi:10.1016/j.ijpe.2008.07.008
31. Tang, O., Nurmaya Musa, S.: Identifying risk issues and research advancements in supply chain risk management. Int. J. Prod. Econ. **133**, 25–34 (2011). doi:10.1016/j.ijpe.2010.06.013
32. Vilko, J.P.P., Hallikas, J.M.: Risk assessment in multimodal supply chains. Int. J. Prod. Econ. **140**, 586–595 (2012). doi:10.1016/j.ijpe.2011.09.010
33. Nakashima, K., Gupta, S.M.: A study on the risk management of multi Kanban system in a closed loop supply chain. Int. J. Prod. Econ. **139**, 65–68 (2012). doi:10.1016/j.ijpe.2012.03. 016
34. Xia, D., Chen, B.: A comprehensive decision-making model for risk management of supply chain. Expert Syst. Appl. **38**, 4957–4966 (2011). doi:10.1016/j.eswa.2010.09.156
35. Giannakis, M., Louis, M.: A multi-agent based framework for supply chain risk management. J. Purch. Supply. Manag. **17**, 23–31 (2011). doi:10.1016/j.pursup.2010.05.001
36. Sofyalıoğlu, Ç., Kartal, B.: The selection of global supply chain risk management strategies by using fuzzy analytical hierarchy process—a case from Turkey. Procedia Soc. Behav. Sci. **58**, 1448–1457 (2012). doi:10.1016/j.sbspro.2012.09.1131
37. Elmsalmi, M., Hachicha, W.: Risks prioritization in global supply networks using MICMAC method: a real case study. In: 2013 International Conference on Advanced Logistics and Transport ICALT 2013, pp. 394–399. (2013). doi:10.1109/ICAdLT.2013.6568491
38. Berenji, H.R., Anantharaman, R.N.: Supply chain risk management : risk assessment in engineering and manufacturing industries. **2** (2011)
39. Liu, P., Peideliugmailcom, E., Wang, T.: Research on risk evaluation in supply chain based on grey relational method. **3**:28–35 (2008)
40. Zhu, X.: Management the risks of outsourcing: time, quality and correlated costs. Transp. Res. Part E Logist. Transp. Rev. 1–13. (2015). doi:10.1016/j.tre.2015.06.005
41. Alkhatib, S.F., Darlington, R., Nguyen, T.T.: Logistics Service Providers (LSPs) evaluation and selection: literature review and framework development. Strateg. Outsourcing An Int. J. (2015)
42. Gunasekaran, A., Irani, Z., Choy, K.-L., et al.: Performance measures and metrics in outsourcing decisions a review for research and applications. Int. J. Prod. Econ. **161**, 153–166 (2014). doi:10.1016/j.ijpe.2014.12.021
43. Pfohl, H., Gallus, P., Thomas, D.: Interpretive structural modeling of supply chain risks. Int. J. Phys. Distrib. Logist. Manag. **41**, 839–859 (2011). doi:10.1108/09600031111175816
44. Kumar, P., Singh, R.K.: A fuzzy AHP and TOPSIS methodology to evaluate 3PL in a supply chain. J. Model. Manag. **7**, 287–303 (2012). doi:10.1108/17465661211283287
45. State, T., Outsourcing, L.: 2016 Third-Party Logistics Study. (2016)
46. Lee, C.K.M., Ching Yeung, Y., Hong, Z.: An integrated framework for outsourcing risk management. Ind. Manag. Data Syst. **112**, 541–558 (2012). doi:10.1108/02635571211225477
47. Wynstra, F., Spring, M., Schoenherr, T.: Service triads: a research agenda for buyer–supplier–customer triads in business services. J. Oper. Manag. **35**, 1–20 (2015). doi:10.1016/j.jom. 2014.10.002
48. Tummala, R., Schoenherr, T.: Assessing and managing risks using the Supply Chain Risk Management Process (SCRMP). Supply Chain Manag. An Int. J. **16**, 474–483 (2011). doi:10. 1108/13598541111171165
49. ICONTEC, NTC 31000 2011: NTC-ISO 31000. (2011)
50. Nan, J., Huo, J.Z., Liu, H.H.: Supply chain purchasing risk evaluation of manufacturing enterprise based on Fuzzy-AHP method. In: 2nd International Conference on Intelligent Computer Technology Automation, ICICTA 2009, vol. 3, pp. 1001–1005. (2009). doi:10. 1109/ICICTA.2009.707

51. Borghesi , A., Gaudenzi, B.: Risk management. How to Assess, Transfer and Communicate Critical Risks. (2013). doi:10.1016/B978-075066842-2/50007-3
52. Bevilacqua, M., Ciarapica, F.E., Giacchetta, G.: A fuzzy-QFD approach to supplier selection. J. Purch. Supply Manag. **12**, 14–27 (2006). doi:10.1016/j.pursup.2006.02.001
53. Costantino, F., De Minicis, M., González-Prida, V., Crespo, A.: On the use of quality function deployment (QFD) for the identification of risks associated to warranty programs. **6**:4440–4449 (2012)
54. Gento, A.M., Minambres, M.D., Redondo, A., Perez, M.E.: QFD application in a service environment: a new approach in risk management in an university. Oper. Res. **1**, 115–132 (2001). doi:10.1007/BF02936289
55. Moosa, I.A.: Operational Risk Management. (2007). doi:10.1057/9780230591486
56. Feria Dominguez, J.M., et al.: El modelo de Distribución de Pérdidas Agregadas (LDA): una aplicación al riesgo operacional. **41013**:1–25 (2007)

Chapter 11
Assessment of Ergonomic Compatibility on the Selection of Advanced Manufacturing Technology

Aide Maldonado-Macías, Arturo Realyvásquez,
Jorge Luis García-Alcaraz, Giner Alor-Hernández,
Jorge Limón-Romero and Liliana Avelar-Sosa

Abstract This paper proposes the development of an expert system for ergonomic compatibility assessment on the selection of Advanced Manufacturing Technology (AMT). Actual models for AMT assessment neglect Human Factors and Ergonomic (HFE) attributes and present deficiencies such as high time consumption and complexity. This approach proposes a novel axiomatic design methodology under fuzzy environment including two stages: the generation of fuzzy If-Then rules using Mamdani's fuzzy inference system and the development of the system by mean of experts' opinions. A numerical example is presented for the selection of three CNC milling machines using the Weighted Ergonomic Incompatibility Content (*WEIC*). The expert system leads to the selection of the best alternative containing the minimum *WEIC*, as the one that better satisfies ergonomic requirements. Development and application of the system may help provide an easier, faster, single or group ergonomic assessment on AMT selection by promoting safer and more ergonomic workplaces in manufacturing companies.

Keywords Advanced manufacturing technology assessment · Ergonomic incompatibility content · Expert system

A. Maldonado-Macías (✉) · A. Realyvásquez · J.L. García-Alcaraz · L. Avelar-Sosa
Departamento de Ingeniería Industrial y de Manufactura, Universidad Autónoma
de Ciudad Juárez, Del Charro Ave. 450 N, 32310 Ciudad Juarez, Chihuahua, Mexico
e-mail: amaldona@uacj.mx

G. Alor-Hernández
Division of Research and Postgraduate Studies, Instituto Tecnológico de Orizaba,
Av. Oriente 9 no. 852 Col. E. Zapata, CP 94320 Orizaba, Veracruz, Mexico
e-mail: galor@itorizaba.edu.mx

J. Limón-Romero
Universidad Autónoma de Baja California, Transpenisular Highway,
Ensenada-Tijuana 3917, Colonia Playitas, 22860 Ensenada, Baja California, Mexico

© Springer International Publishing AG 2017
G. Alor-Hernández and R. Valencia-García (eds.), *Current Trends
on Knowledge-Based Systems*, Intelligent Systems Reference Library 120,
DOI 10.1007/978-3-319-51905-0_11

11.1 Introduction

Manufacturing companies around the world are requested to be more competitive in order to successfully remain in the market. One of the most significant resources they have used to achieve this purpose is increasing their investments in Advanced Manufacturing Technology (AMT). AMT is defined as an umbrella term to describe a variety of technologies that utilize the computers in the manufacturing activities either directly or indirectly. AMT assessment is considered a complex problem since it involves tangible attributes such as production rate, speed, delivery time, process inventory, and machine time availability, among others [1]. On the other hand, intangible attributes concern flexibility, and quality, for instance [2], while tangible and intangible Human Factors and Ergonomics (HFE) attributes have been neglected or obviated in actual models due to the lack of timely and appropriate information [3]. This background shows that HFE have not been considered properly on the design, assessment, and selection of AMT. HFE consider the human capabilities and limitations and integrate them into the design, assessment, selection, and implementation of AMT [4]. As a result, decision makers are not aware of its benefits. In addition, failing to consider ergonomic attributes in the design, assessment, and selection of AMT may cause greater investment in training, higher rates of errors, lower production levels, and poor quality in manufacturing companies [5]. Also, it can generate injuries and accidents which cause severe economic problems for companies that face them [4]. These facts reflect the need to carry out a research that promotes the inclusion of ergonomic attributes for the assessment and selection of AMT by using a more effective and complete approach.

In a fuzzy expert system, information comes from human experts who apply their knowledge about the system on a natural language and allows for the transformation of human basis knowledge into mathematical models [4]. This was one of the reasons to develop a fuzzy expert system approach in this project, which also includes opinions of experts regarding the ergonomic compatibility assessment of AMT. The main objective of this paper is to develop an Expert system (ES) for the assessment of AMT applying a novel axiomatic design methodology under fuzzy environment for the generation of fuzzy If-Then rules. Other objectives are to propose a procedure for ES development and validate the system using a numerical example and a sensitive analysis.

11.2 Literature Review

11.2.1 Models for Assessment and Selection of AMT

A diversity of models can be found in literature for justifying, assessing, and selecting AMT. For example, economic models such as discounted cash flow, net present value, and rate of return are some of the most used for justifying investments of AMT, although they are also some of the most limited ones.

In order to include tangible and intangible attributes into the assessment process of AMT, some models based on fuzzy logic have been developed [6]. For instance, an assessment methodology has conducted a fuzzy analysis of discounted cash flow using linguistic assessments for attributes such as flexibility and quality of AMT [7].

Previous methods and models present at least one of the following shortcomings: (1) they require exact measures to assess intangible attributes, or/and (2) they require expert knowledge to be conducted. In addition, none of them takes into account ergonomic attributes.

Some contributions have been made in this regard, and the Ergonomic Compatibility Assessment Model (ECAM) was developed to evaluate AMT from an ergonomic perspective [4]. Unfortunately, the highest deficiency of this model is the great amount of time and effort needed to perform the assessment. Nevertheless, this model has been taken as reference to develop the ES proposed in this paper, especially for the ergonomic compatibility attributes considered and the fuzzy axiomatic design perspective proposed.

Ergonomic attributes for the selection of AMT Human beings have limitations in their interaction with AMT; and these limitations must be considered when selecting AMT. Otherwise, human inefficiencies and equipment downtimes could affect the production time and the performance of manufacturing systems [8]. A modern manufacturing approach centered on HFE may be more effective, especially if it is based on real productivity improvements, economy, technical feasibility, and the capacity and reliability of equipment [8].

Based on this idea, a new model was proposed for the assessment of AMT. For a better comprehension, see [9, 8]. Ergonomic attributes used in this paper are shown in Fig. 11.1. Appendix A, introduces a review regarding the authors who have considered ergonomic attributes for AMT evaluation.

11.2.2 Applications of Fuzzy Logic in Manufacturing

Fuzzy logic aims to model inherent impreciseness present in our natural language. It captures through the process of inference, uncertainty, ambiguity, and complexity of the human cognitive processes [10]. Fuzzy logic is employed to represent and manipulate inferences though the use of fuzzy *if-then* rules, which are based on linguistic variables [11]. Some reasons for using fuzzy logic are when a system is unknown, when parameters are dynamic, and when constraints within the process and the environment are not easily or feasible to model mathematically [10].

Literature shows that applications of fuzzy logic in manufacturing have been focused on process control and optimization, manufacturing cells and machine controls, manufacturing systems flexibility, among others. For a more detailed review of these applications see [12].

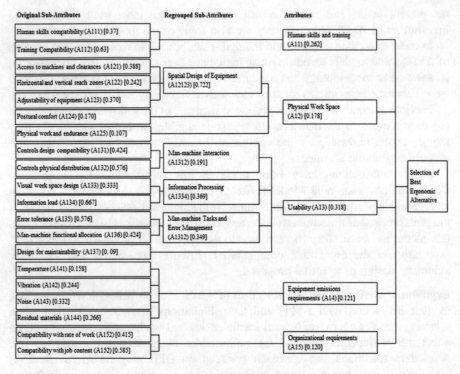

Fig. 11.1 Sub-attributes weights on regrouping

11.2.3 Fuzzy Inference

The process of fuzzy inference consists in formulating an appropriate output space from a fuzzy rule [12]. Fuzzy rules are commonly expressed in a form such as "IF x is A_i THEN y is B_i", where A_i and B_i are fuzzy sets representing fuzzy concepts (linguistic variables) [11]. A number of techniques have been proposed for fuzzy reasoning such as Mamdani's model. According to [12], Mamdani combines fuzzy rules as it is indicated in Eq. (11.1):

$$R(x, y) = \bigvee\nolimits_{i=1}^{n} A_i(x) \wedge B_i(y) \tag{11.1}$$

and for an extended set of rules as indicated in Eq. (11.2) and in Eq. (11.3)

$$R_i : If\ A_{i1}\ and\ A_{i2}\ and \dots and\ A_{ik}\ THEN\ B_i, i = 1, \dots, n \tag{11.2}$$

$$R(x, y) = R(x_1, x_2, .., x_k, y) = \vee_{i=1}^{n}(A_{i1}(x_1) \wedge (A_{i2}(x_2) \wedge \dots \wedge (A_{ik}(x_k) \wedge B_i(y))) \tag{11.3}$$

Fuzzy inference process aims to combine fuzzy rules to produce a fuzzy set (output space). Generally, in most of the applications, a crisp number is desirable; this involves a process named "defuzzification". Defuzzification is the process of formulating a crisp value from a fuzzy set. As well as fuzzy reasoning, there are different methods to carry out the defuzzification process (composite moments and composite maximum). Center of area (COA) or centroid method (composite moments) is one of the most employed techniques; thus, a crisp number is formulated using COA by means of Eq. (11.4):

$$c = \frac{\sum_0^n f(x)_i x_i}{\sum_0^n f(x)_i} \tag{11.4}$$

According to Sivanandam et al. [13], a fuzzy inference system (FIS) consists of a fuzzification interface, a rule base, a database, a decision-making unit, and a defuzzification interface. A FIS with five functional blocks is represented in Fig. 11.2. The function of each block is as follows:

- A rule base containing a number of fuzzy If-Then rules.
- The rule base in the ES corresponds to the set of rules that have been developed.
- A database which defines the membership functions of the fuzzy sets used in the fuzzy rules.
- A decision-making unit that performs the inference operations on the rules.
- A fuzzification interface that transforms crisp inputs into degrees of correspondence with linguistic values.
- A defuzzification interface which transforms the fuzzy results of the inference into a crisp output.
- At the ES, the decision-making can correspond to the person who, based on the rules, takes a final decision on the *EIC* of the ATM.
- The ES of this work, there is not a crisp value at the beginning of the system. Decision-makers introduce linguistic terms to evaluate ergonomic sub-attributes of AMT.

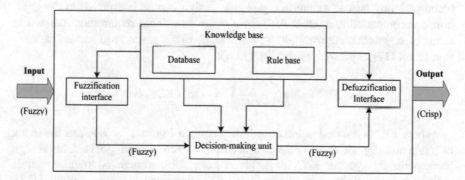

Fig. 11.2 Fuzzy interference system

The ES based on the Mamdani's FIS functions as it follows. First, a number of decision-makers rate the alternatives of AMT according to the degree of ergonomic sub-attributes contained. Ratings are given by means of the linguistic terms. Afterwards, these linguistic terms are defuzzified for every alternative by applying the centroid equation (see Eq. 11.4), and crisp values are obtained for every sub-attribute. Later in this paper, the procedure to develop the expert system will be described.

11.2.4 Axiomatic Design for the Assessment and Selection of AMT

Axiomatic design is a concept that only few authors have combined with fuzzy logic [14]. Axiomatic design theory can be applied to the design systems such as manufacturing systems, software, products, and organizations [15]. The design process comprises four domains: customer domain, functional domain, physical domain, and process domain. In these domains it is specified what the design is intended to do and the methodology to achieve the goal.

Axiomatic design theory has two axioms: the independence axiom and the information axiom. This theory has been applied for ergonomic design. An Ergonomic design is defined as the relation between human capacities and limitations with the manufacturing system requirements [16]. Axiomatic Design Theory axioms that were adapted to ergonomic design goals, as it is stated by Karwowski [15] are:

Independence Axiom: This axiom specifies the need of independence among the compatibility functional requirements, which are defined as the minimum set of independent compatibility requirements that characterize the design goal defined by the ergonomic design parameters.

Human incompatibility Axiom: This axiom specifies the need to minimize the design incompatibility content. Among those design that satisfy the independence axiom, the best design is the one with the lowest incompatibility content. The incompatibility human axiom in ergonomic design can be interpreted as the ergonomic incompatibility content EIC_i for a given functional requirement defined in terms of ergonomic compatibility content ECC_i like a given requirement satisfaction index [17] expressed by the Eq. (11.5):

$$EIC_i = \log_2\left(\frac{1}{ECC_i}\right) = -\log_2 ECC_i \, ints \qquad (11.5)$$

Where *EIC* is the design incompatibility content in terms of *ints*, and the index of compatibility *ECC* is defined according to the design goals. The design incompatibility content *EIC* for a given compatible functional requirement is defined in terms of the compatibility index *ECC* that satisfies this requirement [17].

11.3 Methodology

According to Maldonado et al. [8], the ECEM was developed to assess AMT from an ergonomic perspective. This model was adopted as the basis to develop the ES. According to Celik et al. [18] the linguistic terms Poor (P), Regular (R), Good (G), Very Good (VG), and Excellent (E) were used for intangible attributes and sub-attributes. Also, linguistic terms Very Low (VL), Low (L), Medium (M), High (H), and Very High (VH) were used for tangible attributes and sub-attributes. Table 11.1 shows the linguistic terms and their fuzzy numbers for tangible and intangible sub-attributes. Fuzzy sets for these linguistic terms are shown in [18].

11.3.1 Methods

Methodology developed for the ES was divided into three stages. The first stage involves the formulation of fuzzy rules (If-Then rules), the second stage refers to the development of the ES with Matlab® and its implementation. Steps 2–6 indicated in Stage 1 present the Fuzzy rules formulation, and the consequent for the *EIC* of AMT alternatives is known. Note that a crisp value is obtained in Step 4, and it is then translated into a linguistic term to provide a clearer notion of the *EIC* of AMT. The last stage refers to the selection of the best alternative.

Stage 1: Fuzzy rules formulation

Maldonado et al. [8] proposed different fuzzy sets. These fuzzy sets were adapted to deliver new fuzzy sets to formulate the fuzzy rules. For the tangible attributes five fuzzy sets (Very Low, Low, Medium, High, and Very High) were distributed on a scale with range 0–1. For the intangible attributes the five fuzzy sets Poor, Regular, Good, Very Good, and Excellent were also distributed on a scale with range 0–1. The scale for the Ergonomic Incompatibility Content (*EIC*) was developed based on the fact that membership functions can be assigned to linguistic terms by means of the intuition delivered of the experts' judgment. This scale uses the same linguistic terms than those used for tangible attributes (Very Low, Low, Medium, High, and Very High). Moreover, it comprises the range 0–4 following the range of historical

Table 11.1 Linguistic terms and their fuzzy numbers for tangible sub-attributes

Linguistic term	Fuzzy number	Linguistic term	Fuzzy number
Tangible		Intangible	
Very Low	(0, 0, 0.3)	Poor	(0, 0, 0.3)
Low	(0, 0.25, 0.5)	Regular	(0.2, 0.35, 0.5)
Medium	(0.3, 0.5, 0.7)	Good	(0.4, 0.55, 0.7)
High	(0.5, 0.75, 1)	Very Good	(0.6, 0.75, 0.9)
Very High	(0.7, 1, 1)	Excellent	(0.8, 1, 1)

values obtained in previous cases developed by Maldonado et al. [8] and Maldonado-Macías et al. [26]. Table 11.2 shows the linguistic terms and their fuzzy numbers for the *EIC* scale.

To develop the system it was necessary to decrease the number of fuzzy rules. This was achieved by following the method proposed by Azadeh et al. [19]. The method hierarchically organizes the attributes, classifying into the same group those attributes with common specifications. Once fuzzy sets and linguistic terms were defined, fuzzy rules were derived by following the next steps:

Step 1: Defuzzification

At this step a precise value was associated with each fuzzy set by means of the centroid method applying the Eq. (11.4).

Step 2: Applying the Human Incompatibility Axiom

This axiom states that a design with the minimum human incompatibility content has a greater probability of satisfying ergonomic compatibility requirements. The alternative with the minimum *EIC* is the best ergonomic alternative. This axiom was applied by means of the Eq. (11.6) [17]:

$$EIC_i = \log_2(1/c_i) \tag{11.6}$$

In Eq. (11.6) EIC_i stands for the ergonomic incompatibility content for the attribute i on any alternative, and c_i is the centroid value—compatibility content—for the linguistic term given to the attribute i on the defined alternative. This step is only applicable to sub-attributes.

Step 3: Multiplying by the attributes' weights

Weights of the attributes were normalized after the regrouping according to specifications proposed by Corlett and Clark [20]. For example, sub-attributes A121, A122 and A123 were classified into the group Equipment Spatial Design (A12123), which is in an intermediate level between attributes and sub-attributes. Original weights of attributes can be found in [8]. Once the new weights were defined for each new group of sub-attributes, the EIC of each sub-attribute was multiplied by its corresponding weight.

Step 4: Sum of the Weighted Ergonomic Incompatibility Content (WEIC)

All weighted *EIC* were added in order to get a total *WEIC* for every main attribute. This step is applicable to all sub-attributes, considering all possible combinations of qualifications to the sub-attributes.

Table 11.2 Linguistic terms and their fuzzy numbers for the *EIC* scale

Linguistic term	Fuzzy number
Very low	(0, 0, 0.7)
Low	(0.5, 0.75, 1)
Medium	(0.8, 1.025, 1.25)
High	(1.15, 1.575, 2)
Very high	(1.5, 2.1, 4, 4)

Step 5: Finding the consequent for the attribute of the following hierarchical level

The consequent (linguistic term) for attributes located at the following hierarchical level was derived by applying the Mamdani's fuzzy inference system. The W*EIC* value computed at Step 4 was located in the fuzzy set shown in Fig. 11.3. The ES utilizes *if* a function to determine the *EIC* value sets regions and find the correspondent linguistic term. For instance, in the *EIC* scale, the ES finds a value of 1.79. This value has the highest membership function corresponding to the set of Very High. Then, the following function can be created to define the region were *EIC* is Very High:

If *EIC* ≥ 1.79, THEN *EIC* is Very High.

With the same reasoning, the ES defines regions to the other linguistic terms, and the general function is as follows. In this case (1.79, 1.21, 0.89, 0.55) are the cut values that define the membership of *WEIC* values among the fuzzy sets.

If EIC ≥ 1.79 ⇒ Very High
If 1.21 ≤ EIC < 1.79 ⇒ High
If 0.89 ≤ EIC < 1.21 ⇒ Medium
If 0.55 ≤ EIC < 0.89 ⇒ Low
If EIC < 0.55 ⇒ Very Low

Step 6: Fuzzy rules formulation

Now that the linguistic term for the sub-attributes and the consequent for the attributes are determined, the rule can be formulated as: IF *x* is *A* and *y* is *B*, THEN *z* is *C*. All possible combinations were taken into account. In order to decrease the number of fuzzy rules, some of these were summarized in one rule. For example, consider the following rules:

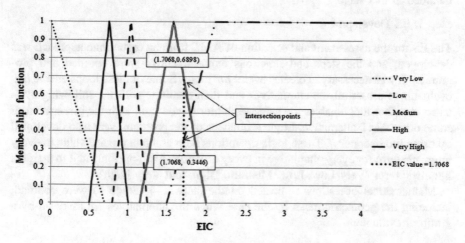

Fig. 11.3 Determining the consequent for an *EIC* equal to 1.7068

IF A111 is Poor AND A112 is Poor, THEN A11 is Very High
IF A111 is Poor AND A112 is Regular, THEN A11 is Very High
IF A111 is Poor AND A112 is Good, THEN A11 is Very High
IF A111 is Poor AND A112 is Very Good, THEN A11 is Very High
IF A111 is Poor AND A112 is Excellent, THEN A11 is Very High

When the antecedent has no effect on the consequent, some rules can be stated in only one rule with no change in the fuzzy rule consequent: IF*A111* is *Poor*, THEN *A11* is *Very High*. Linguistic terms *at least* and *at most* were used to summarize group of rules. For example, the following rules for the attribute *A11*:

IF A111 is Poor AND A112 is Poor, THEN A11 is Very High
IF A111 is Poor AND A112 is Regular, THEN A11 is Very High
IF A111 is Poor AND A112 is Good, THEN A11 is Very High
IF A111 is Poor AND A112 is Very Good, THEN A11 is High
IF A111 is Poor AND A112 is Excellent, THEN A11 is High

They can be summarized in only two rules:

IF *A111* is *Poor* AND *A112* is at most *Good*, THEN *A11* is *Very High*
IF *A111* is *Poor* AND *A112* is at least *Very Good*, THEN *A11* is *High*

As observed, the first of these two final rules (at most *Good*) comprises the first three rules (*Poor*, *Regular*, and *Good*), and the second rule (at least *Very Good*) comprises the two final rules (*Very Good* and *Excellent*).

Stage 2: Development and implementation of the ES

The ES development process was performed using Matlab® 2010 and its functions were defined. Afterwards, the procedure for selecting the best ergonomic alternative is conducted. Also, methods applied for the implementation of the system are included in this stage.

Step 1: ES Development

The ES for the assessment and selection of AMT from an ergonomic approach was developed, and the steps and functions exposed above were codified and programed. It was necessary to create several *for* loops. Some of these loops enable one evaluator to assess all the alternatives and then continue with the following evaluator. Another loop enables the same evaluator to continue with the assessment of some other AMT alternative. Logic sequences of the program were also developed by using the operator *if*. These logic operations help assign a precise value to every linguistic term for every alternative in every ergonomic sub-attribute and to assign a linguistic term (Very Low, Low, Medium, High, and Very High).

Mathematical operations indicated on the steps 1–4 (Stage 1) were codified, including the geometric mean by the case when the alternatives are assessed by a group of evaluators.

Step 2: ES Implementation

ESs can sometimes be implemented against historical results regardless the number of cases against which the system is validated[19–25] ES proposed in this research is validated against the results formulated on three cases study developed by Maldonado et al. [8] and Maldonado-Macías et al. [26] This paper only presents the case study of the milling machines.

Stage 3: Selection of the best alternative

At this stage the expert system offers the resulting values of EIC for each alternative using linguistic terms and crisp values, the minimum EI value is chosen as the best alternative. This alternative is the one that better satisfies ergonomic compatibility requirements for AMT.

11.3.2 Mathematical Model

Let A_i $i = 1...$ n, be the attributes that allow for the measurement of ergonomic incompatibility content (*EIC*) of AMT. Each A_i takes linguistic variations LA_i^j, $j = 1, ..., k$, and it is defined in U_i. LA_i^j are fuzzy sets defined as indicated by Eq. (11.7).

$$LA_i^j = \left\{ \left(e_I, \mu_{LA_i^j}(e_i) \right) : e_i \in U_i \right\} \tag{11.7}$$

Ergonomic incompatibility content *EIC* of an AMT, defined in $U_1 \times U_2 \times ...U_n$, is given by Eq. (11.8).

$$G = f(A_1, A_2, ..., A_n) \tag{11.8}$$

In this equation, f represents the fuzzy or approximate reasoning procedure. In fuzzy logic terminology, the relation between ergonomic compatibility and its observable parameters is expressed by fuzzy as it is stated by Eq. (11.9):

$$\text{IF } LA_1^j \text{ AND } LA_2^j ... \text{AND } LA_n^j \text{ THEN } LEIC \tag{11.9}$$

In this case, *LEIC* is the set of linguistic variations of ergonomic incompatibility content. The rule in Eq. (11.9) can be rewritten as a relation equation in the membership functions domain as it indicated in Eq. (11.10), with the equivalence indicated in Eq. (11.11):

$$\mu_R(e_1, ..., e_n, 1) = [1 \rightarrow [\mu_{AND}(e_1, ..., e_n), \mu_{LEIC}(g)] \tag{11.10}$$

where g $\rightarrow$ represents the fuzzy implication and

$$\mu_{AND}(e_1, \ldots, e_n) = \mu_{LA_1^j}(e_1) \text{AND} \ldots \text{AND} \mu_{LA_n^m}(e_n) \tag{11.11}$$

The relation Eq. (11.10) is the mathematical interpretation of Eq. (11.11). One should select appropriate 'AND' and implication operators for a given context. For instance, if μ_A AND $\mu_B = \mu_A \wedge \mu_B$, then Eq. (11.11) becomes Eq. (11.12):

$$\mu_R(e_1, \ldots, e_n, g) = \vee \cup \left[\mu_{LA_i^j}(e_1) \wedge \ldots \ldots \wedge \mu_{LA_n^m}(e_1), \mu_{LG}(g) \right] \tag{11.12}$$

For random values of LA_1^j, denoted by $LA_1^{j^*}$ (generally $LA_1^j \neq LA_1^{j^*}$), the membership of the fuzzy set LG^*: "Assessment of ergonomic incompatibility content of AMT" is computed by means of Eq. (11.13), where '$\Diamond$' represents the inference.

$$\mu_{LG}(g)* = \mu_{LA_i^{j^*}} \Diamond \mu_{AND}(e_1, \ldots, e_n, g). \tag{11.13}$$

11.3.3 Results Using a Numerical Example

Stage 1: Fuzzy rules formulation

Following subsections present in a detailed way the results obtained from the methodology applied in this research for the fuzzy rules formulation and expert system development using a numerical example on the selection of three milling machines, eight experts were invited to participate in this case study.

Step 1: Defuzzification of the linguistic terms

The following precise values (centroids) for every of the linguistic terms resulted from step 1 described in the methodology section. Table 11.3, show the linguistic terms centroids for tangible and intangible attributes.

Table 11.3 Linguistic terms for tangible and intangible attributes and their centroids

Linguistic term	Centroid	Linguistic term	Centroid
Tangible		Intangible	
Very low	0.9167	Poor	0.0833
Low	0.75	Regular	0.35
Medium	0.5	Good	0.55
High	0.25	Very good	0.75
Very high	0.0833	Excellent	0.95

Step 2: Application of the Human Incompatibility Axiom

Equation (11.6) was applied at this step for each one of the centroids shown in Table 3.3. The *EIC*s for the linguistic terms were derived with this equation. Table 11.4 shows the *EIC* for every linguistic term.

Step 3: Multiplying by the attributes weights

The new weights of sub-attributes after the regrouping are show in brackets in Fig. 11.1. Note that in every group the sum of the weights is 1.

The *EIC* is multiplied by these new weights to get the *WEIC*. For instance, suppose that for the sub-attribute A11 we got all the *EIC*s presented in Table 11.5, then multiplying these *EIC*s for the corresponding weight we obtain the *WEIC*s presented in Table 11.6.

Step 4–5: Results for these steps are included in the numerical example.

Table 11.4 EIC for each linguistic term

Linguistic term	EIC	Linguistic term	EIC
Very low	0.125	Poor	3.585
Low	0.415	Regular	1.514
Medium	1	Good	0.862
High	2	Very good	0.415
Very high	3.585	Excellent	0.074

Table 11.5 *WEIC* for each linguistic term for the sub-attribute A11

Linguistic term	*WEIC*	Linguistic term	*WEIC*
Very low	0.033	Poor	0.939
Low	0.109	Regular	0.397
Medium	0.262	Good	0.226
High	0.524	Very good	0.109
Very high	0.939	Excellent	0.019

Table 11.6 Sample of fuzzy rules for AMT ergonomic compatibility

1. IF A111 is Poor AND A112 is at most Good, THEN A11 is Very High

2. IF A121 is Poor AND A122 is Poor, THEN A12 is Very High

3. IF A131 is Poor, AND A132 is at most Good, AND A134 is Poor, AND A135 is Poor, THEN A13 is Very High

4. IF A141 is High, AND A142 is Very Low, AND A143 is Very Low, AND A144 is at most Low, THEN A14 is Very Low

5. IF A151 is Very Good, AND A152 is at least Very Good, THEN A15 is Very Low

6. IF A11 is Very Low, AND A12 is Very High, AND A13 is Very High, AND A14 is at least Low, AND A15 is High, THEN *EIC* is Very High

Step 6: Fuzzy rules formulation for the assessment of AMT

Most of the fuzzy rules are summarized, although due to the large number them only a few are presented in this chapter. Table 11.6, shows a sample of fuzzy rules for each one of the sub-attributes and a rule for the final *EIC*. Fuzzy rules 1–5 correspond to attributes A11, A12, A13, A14, and A15, respectively. Fuzzy rules 6 shows the final EIC based on the EICs of the attributes. Summarized rules were used to validate the system by means of a sensitivity analysis.

Stage 2: Results of ES Design and Development Operation of the Inference System

A computing program was developed by following the methodology presented in this work to apply the ES on the assessment and selection of AMT from an ergonomic approach. The software Matlab® 2010 was used to develop such program, which allows for the assessment of several alternatives by either one or more evaluators.

During the assessment process, the program enables to identify the current evaluator. Moreover, it provides a description of every attribute. The evaluators assess each sub-attribute for each alternative by giving a linguistic rate. When an evaluator finishes assessing all the alternatives for all sub-attributes, the program automatically grants the turn to the next evaluator.

The program provides the *EIC* for every alternative in every attribute (*A11*, *A12*, *A13*, *A14*, and *A15*), and its corresponding sub-attributes in both numbers and linguistic terms. Finally, it provides the total *EIC* for each alternative and indicates the best alternative from an ergonomic approach.

Appendix B shows the rates assigned by the experts to on a case study of milling machines alternatives. In this table, alternatives are indicated by letters X, Y, and Z, while experts are represented by E1–E8. In addition, rates were abbreviated as it follows: *P = Poor, R = Regular, G = Good, VG = Very Good, E = Excellent, VL = Very Low, L = Low, M = Medium, H = High, and VH = Very High*. In this example, eight experts evaluated three milling machines alternatives, and rates assigned by the experts were introduced in the Matlab® 2010 program. When there are several evaluators, as in this example, the ES applies a geometric mean to the centroid of the rates of a specific sub-attribute for every alternative.

Stage 3: Selection of the best ergonomic alternative

Once the evaluators have assigned a rate to every sub-attribute of the model and for a specific number of alternatives, the ES must provide a final result of EIC. This result is expressed in linguistic terms for every alternative, which results in several cases. In the first case, the ES gives a solution by using linguistic term. For instance, rates were formulated for a set or three alternatives X is equal to Low, Y to High, and Z to Very High logistic term respectively. In this case, it is easy to find the best alternative since all of them have different linguistic ratings. Thus, alternative X is the best alternative since it has the minimum *EIC* value.

Table 11.7 *EIC* for CIM alternatives

Alternative	*EIC* with the ES
X	0.8722
Y	1.1212
Z	0.9615

Sometimes, decision making can be more complex. For instance, equality on the final linguistic term for some alternatives, it can be solved finding the alternative with the minimum value of the WEIC as the best alternative solution.

At the end, the ES provides the final *EIC* for every alternative. Table 11.7, shows the results of the ES in the case study of milling machines. In this case study alternative X was the best alternative since it had the minimum value of *EIC*. This alternative can better satisfy ergonomic compatibility requirements of AMT.

11.3.4 Conclusions and Future Research

An original approach was proposed using Fuzzy Axiomatic Design Rules formulation to develop an ES that can be used to perform group decision-making dealing with tangible and intangible attributes. Based on the validation and the example, it is concluded that the ES efficiently and effectively supports the assessment of AMT from an ergonomic approach, since it allows users to make a more complete decision about this technology including ergonomic compatibility attributes, and by saving time and effort when they compute the outcomes.

Similarly, it is concluded that fuzzy logic offers advantages for the evaluation of tangible and intangible attributes, particularly HFE attributes, when uncertainty and vagueness are present. Moreover, it facilitates gathering and handling opinions from experts. Fuzzy rules generation also offers a fast, easy, and complete means for decision makers to perform their duties by including HFE attributes. They can also focus on which alternative would better satisfy HFE requirements and could be implemented according to their needs. As for the ES, it shows that HFE attributes can be handled effectively including group decision-making processes.

In addition, this ES can also be used to evaluate AMT from an ergonomic perspective and find ergonomic deficiencies in those AMT alternatives with high WEIC. This can help guide the ergonomic design of AMT in order to be more ergonomically compatible. Similarly, future research can lead us to develop an ES that may contribute to evaluate AMT using a holistic approach (i.e. assessment of ergonomics, flexibility, productivity, etc.).

Finally, certain recommendations can be proposed to improve the ES by expanding its application and effective use. It is suggested that the ES be available online for easy access to AMT assessment around the world. Another suggestion is to include other approaches (quality, productivity, flexibility, etc.) into AMT assessment as a holistic focus for decision-making on AMT. It is advised that the

ES be available in several languages. This ES can currently contribute with AMT assessment into a Lean Manufacturing theoretical framework implementation in future companies.

Acknowledgements Authors wish to acknowledge all the reviewers who helped improve this paper through their valuable suggestions and comments. Similarly, we are sincerely thankful to the experts in Advanced Manufacturing Technology and Human Factors and Ergonomics from CENALTEC—National Center of High Technology of Ciudad Juarez, the Autonomous University of Ciudad Juarez, FESTO-Juarez automation, and the Technological Institute of Ciudad Juarez for their priceless contribution of knowledge in this project.

References

1. O'kane, J., Spenceley, J., Taylor, R.: Simulation as an essential tool for advanced manufacturing technology problems. J. Mater. Process. Technol. **107**(1), 412–424 (2000)
2. Mohanty, R., Deshmukh, S.: Advanced manufacturing technology selection: a strategic model for learning and evaluation. Int. J. Prod. Econ. **55**(3), 295–307 (1998)
3. Ordoobadi, S.M., Mulvaney, N.J.: Development of a justification tool for advanced manufacturing technologies: system-wide benefits value analysis. J. Eng. Tech. Manage. **18** (2), 157–184 (2001). doi:10.1016/S0923-4748(01)00033-9
4. Maldonado, A.: Modelo de Evaluación Ergonómica para la Planeación y Selección de Tecnología de Manufactura Avanzada. Instituto Tecnológico de Ciudad Juárez, Disertación Doctoral (2009)
5. Bandrés, M.: Ergonomía. 20 preguntas para aplicar la Ergonomía en la empresa, Segunda edición, p. 25. MAPFRE, Madrid (2001)
6. Chuu, S.-J.: Selecting the advanced manufacturing technology using fuzzy multiple attributes group decision making with multiple fuzzy information. Comput. Ind. Eng. **57**(3), 1033–1042 (2009). doi:10.1016/j.cie.2009.04.011
7. Ertugrul Karsak, E., Tolga, E.: Fuzzy multi-criteria decision-making procedure for evaluating advanced manufacturing system investments. Int. J. Prod. Econ. **69**(1), 49–64 (2001). doi:10. 1016/S0925-5273(00)00081-5
8. Maldonado, A., García, J.L., Alvarado, A., Balderrama, C.O.: A hierarchical fuzzy axiomatic design methodology for ergonomic compatibility evaluation of advanced manufacturing technology. Int. J. Adv. Manuf. Technol. **66**(1–4), 171–186 (2013)
9. Maldonado-Macías, A., Guillén-Anaya, L., Barrón-Díaz, L., García-Alcaraz, L.: Evaluación Ergonómica para la Selección de Tecnología de Manufactura Avanzada: una Propuesta de Software
10. He, W., Zhang, Y., Lee, K., Fuh, J., Nee, A.: Automated process parameter resetting for injection moulding: a fuzzy-neuro approach. J. Intell. Manuf. **9**(1), 17–27 (1998)
11. Prasad, N.R., Walker, C.L. Walker, E.A.: A First Course in Fuzzy and Neural Control: CHAPMAN and HALL/CRC-2003 (2003)
12. Azadegan, A., Porobic, L., Ghazinoory, S., Samouei, P., Saman Kheirkhah, A.: Fuzzy logic in manufacturing: a review of literature and a specialized application. Int. J. Prod. Econ. **132**(2), 258–270 (2011). doi:10.1016/j.ijpe.2011.04.018
13. Sivanandam, S., Sumathi, S., Deepa, S.: Introduction to Fuzzy Logic Using Matlab, vol. 1. Springer (2007)
14. Iancu, I.: *A Mamdani type fuzzy logic controller*: INTECH Open Access Publisher (2012)
15. Kahraman, C., Çebı̇´, S.: A new multi-attribute decision making method: hierarchical fuzzy axiomatic design. Expert Syst. Appl. **36**(3, Part 1), 4848–4861 (2009). doi:10.1016/j.eswa. 2008.05.041

16. Suh, N.P.: Axiomatic design theory for systems. Res. Eng. Design **10**(4), 189–209 (1998)
17. Karwowski, W.: Ergonomics and human factors: the paradigms for science, engineering, design, technology and management of human-compatible systems. Ergonomics **48**(5), 436–463 (2005)
18. Celik, M., Kahraman, C., Cebi, S., Er, I.D.: Fuzzy axiomatic design-based performance evaluation model for docking facilities in shipbuilding industry: the case of Turkish shipyards. Expert Syst. Appl. **36**(1), 599–615 (2009)
19. Azadeh, A., Fam, I.M., Khoshnoud, M., Nikafrouz, M.: Design and implementation of a fuzzy expert system for performance assessment of an integrated health, safety, environment (HSE) and ergonomics system: the case of a gas refinery. Inf. Sci. **178**(22), 4280–4300 (2008). doi:10.1016/j.ins.2008.06.026
20. Corlet, E., Clark, T.: The ergonomics of workspaces and machines. Taylor & Francis, Bristol (1995)
21. Atalay, K.D., Eraslan, E.: Multi-criteria usability evaluation of electronic devices in a fuzzy environment. Human Factors Ergon. Manuf. Serv. Ind. **24**(3), 336–347 (2014)
22. Balogh, I., Ohlsson, K., Hansson, G.-Å., Engström, T., Skerfving, S.: Increasing the degree of automation in a production system: consequences for the physical workload. Int. J. Ind. Ergon. **36**(4), 353–365 (2006)
23. Battini, D., Faccio, M., Persona, A., Sgarbossa, F.: New methodological framework to improve productivity and ergonomics in assembly system design. Int. J. Ind. Ergon. **41**(1), 30–42 (2011)
24. Battini, D., Persona, A., Sgarbossa, F.: Innovative real-time system to integrate ergonomic evaluations into warehouse design and management. Comput. Ind. Eng. **77**, 1–10 (2014)
25. Besnard, D., Cacitti, L.: Interface changes causing accidents. An empirical study of negative transfer. Int. J. Hum. Comput. Stud. **62**(1), 105–125 (2005)
26. Maldonado-Macías, A., Alvarado, A., García, J.L., Balderrama, C.O.: Intuitionistic fuzzy TOPSIS for ergonomic compatibility evaluation of advanced manufacturing technology. Int. J. Adv. Manuf. Technol. **70**(9–12), 2283–2292 (2014)

Chapter 12
Developing Geo-recommender Systems for Industry

Edith Verdejo-Palacios, Giner Alor-Hernández, Cuauhtémoc Sánchez-Ramírez, Lisbeth Rodríguez-Mazahua, José Luis Sánchez-Cervantes and Susana Itzel Pérez-Rodríguez

Abstract Recommender systems are broadly used to provide filtered information from a large amount of elements. They provide personalized recommendations on products or services to users. The recommendations are intended to provide interesting elements to users. Nowadays, recommender systems and geolocation services have focused the attention of many users, this attention has produced a new kind of recommender system called Geo-recommender. Geo-recommender systems can successfully suggest different places depending on the users' interest and current location. This characteristic is useful in market competition, since it allows a better analysis of the study of business locations. Geolocation is key factor to obtain the desired business success; while a business is closer to customers, the benefits are greater. In this chapter, we propose an integration architecture for developing geo-recommender systems for industry. Different case studies are described where the use of geo-recommender systems has taken relevance. The architecture proposed has a layered design, where the functionalities and interrelations of the layer components are distributed in order to ensure maintenance and scalability. Also, a geo-recommender system prototype called GEOREMSYS was developed as a

E. Verdejo-Palacios · G. Alor-Hernández (✉) · C. Sánchez-Ramírez · L. Rodríguez-Mazahua
Division of Research and Postgraduate Studies, Instituto Tecnológico de Orizaba,
Av. Oriente 9 no. 852 Col. E. Zapata, CP 94320 Orizaba, Veracruz, Mexico
e-mail: galor@itorizaba.edu.mx

E. Verdejo-Palacios
e-mail: everdejo@acm.org

C. Sánchez-Ramírez
e-mail: csanchez@itorizaba.edu.mx

L. Rodríguez-Mazahua
e-mail: lrodriguez@itorizaba.edu.mx

J.L. Sánchez-Cervantes
CONACYT-Instituto Tecnológico de Orizaba, Orizaba, Veracruz, Mexico
e-mail: jsanchezc@ito-depi.edu.mx

S.I. Pérez-Rodríguez
Universidad Popular Autónoma Del Estado de Puebla, Puebla, Mexico
e-mail: susanap@gmail.com

© Springer International Publishing AG 2017
G. Alor-Hernández and R. Valencia-García (eds.), *Current Trends on Knowledge-Based Systems*, Intelligent Systems Reference Library 120, DOI 10.1007/978-3-319-51905-0_12

241

proof of concept of the architecture proposed. GEOREMSYS uses collaborative filtering techniques by giving possible locations for Point of Sale (POS).

12.1 Introduction

Geo-recommender systems represent a new kind of recommender systems. They are able to produce recommendations based on the geographical location of both users and businesses [1]. The main benefit of geo-recommender systems is that they combine the characteristics of a recommender system with the features of a Geographic Information System (GIS), mainly for the context of leisure activities [1, 2]. In industry, enterprises require to know the best geographical location to establish a POS. A POS is the physical location at which goods are sold to customers [3]. To achieve this activity, the use of GIS through geographic models represents a new field of research from a commercial perspective, which is a topic not enough exploited. However, studies that have addressed this topic fail to integrate context, current time and user needs in order to generate recommendations based on the emerging behaviours of both users' and customer's needs. To address this gap, geo-recommender systems provide personalized recommendations to users through analysis techniques for studying user's behaviour. Currently, there is a need for integrating geographical information into traditional recommender systems. This could help to the commercial sector in order to improve its services, since it is possible to obtain an (best) option between many options [4–6]; i.e. recommender systems can provide personalized recommendations to users, among all the large information amount through behaviour analysis techniques. However, they fail to provide suggestions based on the user's physical location. Therefore, geo-recommender systems, being the combination of recommender systems with GIS, are a successful alternative to provide recommendations based on demographic and economic characteristics. Based on this understanding, this chapter discusses different perspectives on the importance of applying geographical information to recommendations. Similarly, we propose a prototype architecture of a geo-recommender system that aims at facing the current issues with traditional recommender systems.

This paper is structured as follows. Section 12.2 discusses the state-of-the-art on geo-recommender systems and GIS. Section 12.3 introduces a software architecture to develop geo-recommender systems. Section 12.4 presents five case studies as a proof of concept of the architecture proposed. Finally, Sect. 12.5 emphasizes conclusions.

12.2 State of the Art

This section presents a compilation of related works about geo-recommender systems and GIS.

12.2.1 *Recommender Systems in Different Domains*

Batet et al. [7] developed an agent-based recommender system for mobile devices, which produced recommendations on interesting leisure activities that were close to the user's physical location. Similarly, Noguera et al. [2] pointed out that current change in e-tourism required services offering tourists relevant information to choose POIs (Point of Interest) according to their physical location and preferences. Therefore, the authors proposed a mobile 3D hybrid recommender system for Spanish restaurants located in the province of Jaén. From a similar perspective, Li et al. [8] created a group-coupon recommender mechanism in order to promote location-sensitive products. Authors used a tree-like structure to categorize products and used data in order to construct the user network and gather users' behaviour data. Colombo et al. [1] discussed a context-aware hybrid mobile recommender system for movie showtimes. The system discards movies that are no longer in movie theaters and generates a list of movie showtimes that the user is likely to attend based on his/her preferences, the distances between the user's current position and each one of the movie theaters, and the time required to get to each destination. Likewise, Yu et al. [9] introduced a mobile inference mechanism grounded in location-based service and knowledge. In addition, Pliakos and Kotropoulos [10] created a recommendation method for touristic POIs (Point of Interest) based on images retrieved from social networks through location services. By constructing a method with image attributes and geographical information, authors concluded that the best touristic recommendations were obtained using latent semantic indexing and image classification and location.

Yin et al. [11] developed a travel path search system from geo-tagged photos retrieved from a social network. The system shows the place visited from previous tourists and provides information on when it was visited. Similarly, the system measures similarity of two paths to facilitate trip planning. Also, Fu et al. [12] addressed the need for a travel planning system. The system has a traveling database composed of photos of popular attractions retrieved from TripAdvisor and Yahoo!Travel, and it analyzes these pictures in order to obtain the attraction's travel information, such as popularity, usual stay time, daily visiting times, and visual scenes. This recommender mechanism was developed by using an algorithm that inserts places to visit and the duration of the stay considering the user's interests and the relationship among destinations. Finally, Wei et al. [13] argued that the popularity of a given POI (Point of Interest) was related to certain characteristics of the people who visited it. Therefore, authors analyzed the features of a group of people (closeness and size) obtained from geo-social networking data, and they showed that the best recommendations considered the behaviour of the group.

López-Ornelas et al. [14] proposed a geo-collaborative mobile prototype as a tool for encouraging urban mobility in Mexico City. The prototype relies on collaborative filtering and aims at supporting pedestrians' decision making regarding the best route to get to a place. Schedl and Schnitzer [15] developed a hybrid music recommendation algorithms based on geo-spatial integration of similarities between the user's preferences and the music content. Experiments conducted indicated that the hybrid recommendation approaches improved traditional recommender mechanisms. Finally, in their work, Huang et al. [16] addressed the issue of urban mobility in POI recommendation. They proposed a model based on a factor graph model, the integration of geographical distribution, and user's behaviour in order to provide appropriate POIs that satisfy users' interests. The authors concluded that combining geographical information with user's social information offered better recommendation results than other POI recommendation mechanisms.

12.2.2 Geographic Information Systems Applied to Environmental and Urban Studies

Castro et al. [17] employed a traditional mathematical model used in epidemiology to understand, model, and analyze through system dynamics the critical factors in the propagation of epidemics. They employed a geographic information system (GIS) to visualize and interpret the fluctuations of healthy, infected, and recovered patients. On the other hand, Corner et al. [18] carried out a study in order to identify the effects of waste from fish farms. In this study, authors combined spreadsheets and a GIS through a dispersion module. Also, Vairavamoorthy et al. [19] proposed a GIS-based risk system that predicts the risks of polluted water from sewers, drains, and ditches entering water distribution systems. Likewise, Radiarta et al. [20] carried out a GIS-based multi-criteria evaluation that used satellite data and ground-truth data in order to identify the most suitable places for developing scallop aquaculture in Japan.

Xu and Volker [21] proposed a study of residential areas in urban development. The study proposed a model integrating GIS, system dynamics with 2D and 3D modelling and visualization (GISSD). From a different perspective, Suarez et al. [22] argued that competition among and the performance of franchises were greatly affected by factors related to their location and the quality of facilities. Therefore, authors employed competitive location models and GIS tools. Likewise, Roig et al. [23] developed a methodology for the process of selecting a retail site location. The methodology combined GIS with the Analytical Hierarchy Process (AHP). The former is used to visualize data influencing decision-making process and the AHP, which consists in defining a model through criteria associated with geodemand and geocompetition. Finally, Casillas et al. [24] proposed a method for estimating time and intensity of the Urban Heat Island by interpolating air temperatures generated in a GIS.

Table 12.1 Comparative analysis of the literature

NU = Not used Article	NM = Not mentioned Objective	H = Hybrid Recommender system	CF = Collaborative filtering GIS
Castro et al. [17]	Plan, propose, and develop a model to study epidemics using GIS and system dynamics	NU	ArcGIS
Corner et al. [18]	Analyze, design, and construct a GIS-based module of marine waste distribution	NU	TerrSet (IDRISI)
Vairavamoorthy et al. [19]	Develop a GIS-based software system that predicts risks associated with water distribution systems	NU	ArcGIS
Radiarta et al. [20]	Build a GIS-based multi-criteria evaluation model	NU	ArcGIS
Yu et al. [9]	Develop a recommender system from the construction of collective knowledge	H NM	NU
Xu and Volker [21]	Develop a 3D GISSD model for sustainability assessment of urban residential areas	NU	ArcGIS
Suarez et al. [22]	Design and build optimal location tools and models for franchises	NU	ArcGIS
Batet et al. [7]	Develop an agent-based hybrid mobile recommender system for activities and events that are available in the city	H NM	NU
Noguera et al. [2]	Develop and integrate a mobile 3D-GIS hybrid recommender system that is sensitive to locations	H NM	NM
Roig et al. [23]	Develop a methodology for the process of selecting a retail site location using GIS and AHP	NU	ArcGIS
Casillas et al. [24]	Apply and validate system dynamics in order to estimate time and intensity of the Urban Heat Island	NU	TerrSet (IDRISI)
Li et al. [8]	Develop a group-coupon recommender mechanism in order to promote location-sensitive products	H NM	NU
Colombo et al. [1]	Develop a recommender system for movie showtimes that is sensitive to location, time, and users	BC NM	NU

(continued)

Table 12.1 (continued)

NU = Not used Article	NM = Not mentioned Objective	H = Hybrid Recommender system	CF = Collaborative filtering GIS
Pliakos and Kotropoulos [10]	Build a recommender method for tourist attractions using latent semantic indexing of images	H NU	NU
Yin et al. [11]	Develop a travel path search system from geo-tagged information of previous tourists	H NM	NU
Fu et al. [12]	Develop a travel planning system that inserts activities and the duration of the stays of the places to be visited	H NM	NU
Wei et al. [13]	Analyze the behavior characteristics of groups having in common a given POI	NU	NU
López-Ornelas et al. [14]	Design the prototype of a recommender system for pedestrians	CF NM	NU
Schedl and Schnitzer [15]	Design hybrid music recommendation algorithms based on geo-spatial integration and the music content	H NM	NU
Huang et al. [16]	Design a model based on a factor graph model, the integration of geographical distribution, and user's behavior	H NM	NU

Table 12.1 presents a comparative analysis of the literature reviewed by considering GIS and recommender systems.

The analysis of the works listed above shows the different domains where recommender systems and GIS are used. As can be observed, the commercial domain is the most prominent. Also, it seems that search of POIs is being mainly addressed by GISs, which fail to integrate geographical influence, frequency of check-in for locations and social influence. Therefore, no research has reported a tool that provides a comprehensive and integrated geographical analysis for the process of selecting a POS through recommender mechanisms that can provide satisfactory recommendations.

12.3   How to Develop a Geographic Recommender System?

This section discusses a generic architecture of a recommender system. The architecture components, their functions, as well as the relationships among them are described. Also, note that it is a multi-layered architecture. This design allows for scalability and maintenance, since tasks and responsibilities are distributed along all the layers. Figure 12.1 depicts the general structure of the proposed integration architecture. The function of each layer is explained below:

Presentation layer: This layer acts as the user interface and serves as a means of communication between results obtained in the other layers and the user. In this layer, the user can obtain information about a specific city in order to select a POS, send the location of the place in the form of latitude and longitude, and visualize the final recommendation on a map.

Integration layer: This layer redirects the requests to the services requested in the presentation layer. It also allows for the construction of responses. In this layer, two components are located: (1) Request Analyzer Component that is responsible for identifying all user requests and sending them to the Services layer; (2) Response Builder Component that receives the responses coming from the Services Layer, the component is responsible for processing information and sending it to the user.

Services layer: This layer is responsible for many of the operations that allow the system to function. This layer contains GIS, recommendation, and services interface modules. Similarly, this layer is responsible for generating the recommendation and provides users the necessary information regarding the restrictions for establishing a POS.

Service Provider: This layer contains the entities providing the geolocation service. These entities offer a Web mapping applications service, which returns a physical coordinates (location) and accuracy radius based on information about cell towers and WiFi nodes that the mobile client can detect. Some location-based service providers are OpenStreetMap™ API, Bing™ Maps Geocode, and Google™ Maps. These RESTful Web services return data in the form of JSON or XML-based files.

Data Access layer: This layer searches data and stores them in the database requested by the Service Provider layer. The Data Access layer also allows for encapsulating data through the different entities and executing insert, update, delete, and query operations thanks to the SQL data generator.

Data layer: This layer stores information about a population, including its settlements and establishments (movie theaters, schools, orphanages, retirement houses, hospitals, churches, nurseries, markets, stadiums, auditoriums, and theaters).

In this architecture proposed, every module has a well-defined function, which is described below:

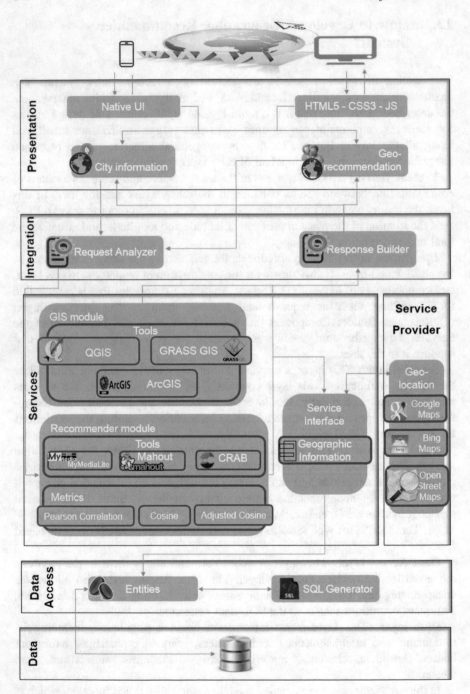

Fig. 12.1 Architecture of a geo-recommender system

GIS module: This module is responsible for building the geographic model based on the features obtained from querying the geographic characteristics of both the place area and the establishments requested to the Service Provider layer. Some tools employed in this module are QGIS, GRASS GIS, and ArcGIS. The first two are free open sources tools, while ArcGIS is available in three license levels. However, the three tools allow for the creation of geographic models from the unification and analysis of geospatial data.

Recommender module: This module is responsible for offering recommendations by correlating the user's profile with other profiles. Likewise, the module makes sure the location of the POS is appropriate.

Recommender systems can be classified into five different categories depending on the technique employed to predict the utility of the items for the user, i.e., according to the recommendation technique:

1. **Content-based recommender systems**: these systems recommend items that are similar to the ones that the user liked in the past.
2. **Collaborative filtering recommender systems**: these systems recommend to the active user the items that other users with similar tastes liked in the past.
3. **Demographic recommender systems**: these systems recommend items based on the identification of the demographic niche the user fits better according to a personal demographic profile.
4. **Knowledge-based recommender systems**: these systems recommend items based on either inferences about user preferences or specific domain knowledge about how items meet user preferences.
5. **Hybrid recommender systems**: these systems are based on the combination of the above mentioned techniques.

Collaborative filtering (CF) is the most suitable technique for developing recommender systems. This technique measures similarity between the different location sites through the following algorithms: Pearson correlation, cosine similarity, and adjusted cosine similarity. It is generally advised to create recommendation models by using different algorithms to calculate similarity, and then, to assess recommendations through recall and precision metrics. Some Collaborative Filtering APIs for recommender systems development are Apache Mahout, MyMediaLite, and CRAB. All of them are open source, and the compatible programming languages are Java and Python; Java for Apache Mahout and Python for MyMediaLite and CRAB.

Service interface: It allows for connecting the location-based and information service providers to the Internet. Also, it helps create complete contents by combining data from multiple Web services.

12.3.1 Development Tools for Recommender Systems

This section discusses different tools for building a recommender system. Mainly, it is necessary to identify which APIs serve to build recommendations and which ones help develop geo-location-based Web applications. Some examples of these tools are Apache Mahout, MyMediaLite™, Weka, and Duine™.

Apache Mahout is an API that allows for solving clustering, classification, and collaborative filtering problems [25]. The Mahout's aim is to work with big volumes of information and distributed systems, and to rely on the community support in order to offer better functions and tackle arising mistakes [26]. Nowadays, Apache Mahout supports the following algorithms described in Table 12.2.

MyMediaLite™ targets at recommender systems based on collaborative filtering [27]. It works under two different scenarios: "rating prediction" and "item prediction from positive-only implicit feedback." The first algorithm estimates unknown ratings from a given set of known ratings and possibly additional data, such as user or item attributes. The predicted ratings can then indicate users how much they will like an item, or the system can suggest items with high predicted ratings. Finally,

Table 12.2 Algorithms in Mahout

Algorithm	Description
Logistic Regression, solved by Stochastic Gradient Descent (SGD)	Classifies text into categories. It is a fast, simple, sequential classifier capable of online learning in demanding environments
Hidden Markov Models (HMM)	Useful in part-of-speech tagging of text; speech recognition. It allows for sequential and parallel implementations of the classic classification algorithm designed to model real-world processes when the underlying generation process is unknown
Singular Value Decomposition (SVD)	Designed to reduce noise in large matrices, thereby making them smaller and easier to work on. Used as a precursor to clustering, recommenders, and classification to do feature selection automatically
Dirichlet clustering	Model-based approach to clustering that determines membership based on whether the data fit into the underlying model. Useful when data have overlap or hierarchy
Spectral clustering	Family of similar approaches that use a graph-based approach to determine cluster membership. Useful for exploring large, unseen data sets
Minhash clustering	Uses a hashing strategy to group similar items together, thereby producing clusters
Numerous recommender improvements	Distributed co-occurrence, SVD, Alternating Least-Squares
Collocations	Map-Reduce enables collocation implementation. Useful for finding statistically interesting phrases in a text

the second algorithm determines the items on which a user will perform a certain action from such past events [28].

Meanwhile, CRAB is a fast recommender engine for Python that integrates classic information filtering recommendation algorithms in the world of scientific Python packages. This framework is useful for systems based on collaborative filtering. CRAB provides a generic interface for recommender systems implementation, among them the collaborative filtering approaches such as User-based and Item-based filtering, which are already available for use [29]. Moreover, CRAB has the following useful segmented features [30]:

- **Recommender Algorithms**: User-Based Filtering and Item-Based Filtering
- **Work in progress**: Slope One, SVD, Evaluation of Recommenders
- **Planned**: Sparse Matrices, REST APIs.

Weka is another tool, which provides a comprehensive collection of machine learning algorithms and data preprocessing tools to researchers and practitioners alike [31]. The process of building a recommender system is the following [32]:

- Create features for every item in your training and testing set.
- Cluster all of the items into a certain number of clusters using XMeans.
- If the user likes a specific item, recommend him/her another item from the same cluster as the first item.

Duine™ is an open-source hybrid recommendation system. It allows users to develop their own prediction engines. Likewise, Duine™ contains a set of recommendation techniques, ways to combine techniques into recommendation strategies, a profile manager, and it allows users to add their own recommender algorithm to the system. It uses switching hybridization method in the selection of prediction techniques [33]. These techniques [34] are listed in the following Table 12.3.

As regards geolocation tools, some of the most common and useful are Google™ Maps, Bing™ Maps, and OpenStreetMap™ (OSM). Google™ Maps offers a maps Web service giving a precise radio location [35]. The service uses the Representational State Transfer (REST) architectural style that uses the four HTTP methods—GET, POST, PUT, and DELETE—to execute different operations [36]. Also, Google™ Maps provides information about geographical regions and sites worldwide with a focus on road and traffic systems. In order to achieve this, Google™ Maps combines aerial satellite imagery with conventional road maps [37]. The Google™ Maps service uses geolocation data to specify the geographical location on a map. The geolocation data used by Google Maps uses geographic position defined by a latitude [−90(S), 90(N) degrees] and longitude [−180(W), 180(E) degrees] coordinate system [38]. Finally, the API also provides an efficient unit for building applications using spatial geographic data and provides GIS features.

Bing™ Maps is a map service similar to GIS tools. It allows for exact lengths and latitudes, maps visualization, and information analysis [39]. Bing™ Maps API is a commercial platform from Microsoft™ that provides a set of geospatial services. The service platform provides data resources and comprehensive APIs to

Table 12.3 Prediction techniques in Duine

Prediction technique	Description
User average	Returns the average rating of the user for all content items that the user has rated in the past
TopN deviation	Returns a prediction based on how all other users have rated this content item in the past
Social filtering	Bases its prediction on how other users having similar interests as the current users have rated the content item
Already known	When the user has already rated the item, that rate is returned, otherwise no prediction is returned
Genre Least Mean Square (LMS)	Bases its prediction on the categories or genres of a piece of information and learns how interested the user is in information of certain categories or genres
Case-based reasoning	Bases its predictions on similar items that the user has already rated in the past
Information filtering	Bases its prediction on a large piece of text describing an item and the known and learned interests of the user by comparing the text with the interests of the user

complement existing GIS and easily build Web mapping interfaces. Bing™ Maps includes modern Web technologies, and it thus brings GIS services to the non-GIS user with easy to use web interfaces, intuitive navigation, and high performance search results [40].

Finally, OpenStreetMap™ (OMS) is a project maintained by a large group of volunteers that create and distribute geographic data for the world [41]. Some features have been implemented in a variety of applications, such as routing 3D modelling, disaster management, and land use mapping. The aim of the OMS project is to create a free editable map of the world. The project maintains a database of geographic elements (nodes, ways, and relations) and features (e.g. streets, buildings, and landmarks). These data are collected and provided by volunteers by using GPS devices, aerial imagery, and local knowledge. In order to get data, OMS gives an API over the HyperText Transfer Protocol (HTTP) for getting raw data and putting them into the OSM database. The main API (currently in version 0.6) has calls to get elements (and all other elements referenced by it) by, among other things, their ID and a bounding box. However, the requests are limited (e.g. currently only an area of 0.25 square degrees can be queried) [42].

12.4 Usage Scenarios of Geographic Recommender Systems

A scenario of use refers to the definition of a problem by means of arguments that allow for its understanding, resulting in the validation of hypotheses. Usage scenarios presented in this chapter are all related to POI (Point of Interest) recommendations in the forms of POS (Points of Sale), shorter routes for product delivery, touristic places, public transportation, and sale offers within a specific radius/distance.

12.4.1 Geo-recommendations for Selecting Points of Sale (Pos)

In this first scenario we present the following premises:

1. A company wishes to establish POS in the Mexican city of Orizaba, Veracruz, in order to market its product.
2. Every POS must comply with current state regulations for the construction, installation, maintenance, and operation of retail fuel stations from Secretariat of the Interior in Mexico (SEGOB).
3. The company wishes to automate the process of finding the appropriate POS locations in a specific city.

How can the company meet the conditions mentioned in number 2 in order to successfully establish its POS? The first step is to define the conditions and constraints for selecting a POS. From this perspective, the state regulations for actions construction, installation, conservation and operation of gas stations in gas station and carburization for retail fuel services stations states that any place destined for retail fuel services stations should be at least 300 meters away from any public institution, such as schools, hospitals, orphanages, nurseries, and retirement houses, and 150 m away from any other establishment, such as movie theaters, markets, stadiums, auditoriums, and churches [43]. These conditions provide the necessary specifications to identify suitable areas for the fuel stations.

As second step, for every establishment to be considered in the process of selecting a POS (e.g. schools, hospitals, supermarkets, among others), its geographic information (latitude and longitude) must be searched by connecting to the Google™ Maps API. To obtain such information, an invocation should be built to INEGI's API DENUE and retrieved data are stored. The National Institute of Geography, Statistics and Informatics (INEGI) is an autonomous agency of the Mexican Government dedicated to coordinate the National System of Statistical and Geographical Information of the country. INEGI offers a wealth of important and helpful information on doing business in Mexico. Likewise, an update mechanism must be developed in order to ensure accuracy of information at all times.

As the third step, through an application, the query service should be offered by using a search form. In this form, the user must enter the desired location for the POS, by indicating latitude and longitude. If the desired location is not suitable (i.e. it does not meet the regulations' specifications), the application notifies that such location is inappropriate. Results from the search are displayed on a map in different colors. Green circles can recommend optimal locations, while orange circles can show mildly appropriate sites, and red circles can indicate inappropriate locations. Also, when the user selects a circle, the application can display further details. It shows how suitable it is to establish a POS in that place and identifies establishments settled around the area (hospitals, retirement houses, nurseries, orphanages, schools, markets, churches, auditoriums, stadiums, and movie theaters).

12.4.2 Geo-recommendations for Product Deliveries

In the second scenario, we propose the following premises:

1. A water purifier company wants to distribute its products. Therefore, it wishes to know the most appropriate delivery routes for deliverers.
2. Deliverers must have a mobile application that allows them to identify the closest customers depending on their physical location.
3. The company wishes to automate the delivery process by selecting the most appropriate routes to distribute the product.

How can the water purifier company satisfy its needs? In order to develop a user (i.e. deliverers) location-sensitive mechanism, the company first needs to have a compendium of all their customers' addresses. Similarly, it is important to connect to the traffic service provided by Google™ Maps API, which displays traffic conditions in real time of major roads. Then, we need to develop a module that measures the shortest distances and construct the geo-recommender system. This results in a mobile application that is able to inform the deliverer of his/her current physical location, the location of the closest customers, as well as the shortest way of reaching them.

12.4.3 Touristic Geo-recommendations

In the third scenario, we present the following premises:

1. The office of tourism in a given city wants to have a touristic application that allows tourists to find the closest and most appropriate hotels according to their preferences.
2. A mobile application that is able to pinpoint all the hotels in the city must be developed.
3. The office of tourism wishes to automate the process of locating hotels.

How can the office of tourism meet the need for a tourism mobile application with such requirements? The first step is to search for the geographic information of every hotel of the city in question. This is achieved by connecting to Google ™Maps API. Afterward, a call needs to be built to INEGI's DENUE API, and retrieved data must be stored. Then, the location-sensitive mechanism must be developed, as well as a questionnaire for the assessment of tourists' behavior. This allows the recommender system to offer recommendations considering both the users' physical location and their personal preferences. This solves the cold-start problem of recommender systems.

12.4.4 Geo-recommendation for Public Transportation

Most people use different means of public transportation in order to reach their destination. The problem is that public transportation has not precise schedules. Thus, waiting and delays are every-day problems that need to be addressed. In this fourth scenario, we present the following premises:

1. The office of public transportation in a city wants to offer citizens an application that provides accurate schedules and the length of the trips based on the user's physical location. The application must be able to provide transportation recommendations considering such information.
2. The application must contain all necessary information about the schedules and routes taken.
3. The application must also display real length of the user's trip and arrival notifications.

How can the office of public transportation offer citizens a mobile application with the aforementioned characteristics? The first step is to know the routes and schedules of buses and the subway. In this case, the office must provide such information. Afterward, it is necessary to both construct a user's location-sensitive mechanism and integrate Google™ Maps API into this mechanism in order to know the traffic conditions. In this integration, it is also necessary to visualize the user's physical location before he/she gets on the bus/train. This would allow the system to recommend trip alternatives depending on the user's location and final destination.

12.4.5 Geo-recommendations for Sale Offers Within a Specific Radius

For the last scenario, we discuss the development of a system that recommends sales offers within a specific radius.

1. The chamber of commerce in a city wishes to create a mobile application that notifies users of sale offers.
2. The application must be capable of identifying a perimeter of interest based on the user's physical location.
3. The application also ought to send notifications of sales offers considering the user's physical location and interests.

How can the chamber of commerce offer this mobile application? First, it is necessary to search data on every business' location and store them in a geo-database. Afterward, locations are placed on a map of the city in question using Google™ Maps API. Then, we need to construct a location-sensitive module that ensures API connection to the GPS of mobile devices in order to recognize the user's current location. Next, we need to develop a business-exclusive component on which every store enters its sale offers. This way, when the location-sensitive component activates as the user walks, the third module is equally activated. The third module generates the recommendation based on the user's location and preferences. A mobile application with these characteristics would have a positive impact on the environment and today's lifestyles.

12.4.6 GEOREMSYS: A Geo-recommender System for Selecting POS

GEOREMSYS has the functionalities described in the first scenario. The objective of our recommender mechanism is to assist users in the process of selecting the appropriate POS. GEOREMSYS is a Web-based application developed on Java Server Faces (JSF) framework and PrimeFaces. JSF is a model-view-presenter framework typically used to create HTML form based web applications. It simplifies development by providing a component-centric approach to developing Java

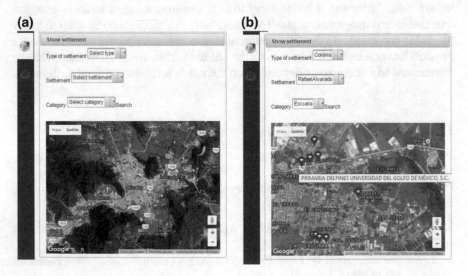

Fig. 12.2 Visualizing establishments for category by using GEOREMSYS. **a** Form before selecting, **b** form after selecting

Web user interfaces. By other hand, PrimeFaces is an open source user interface (UI) component library for JavaServer Faces (JSF) based applications with one jar, zero-configuration and no required dependencies. GEOREMSYS presents the following four functionalities:

First, GEOREMSYS searches establishments (schools, nurseries, theaters, orphanages, retirement houses, hospitals, among others) around a specific settlement or geographic area within a city (suburb, county, among others) by selecting the name of the city, the type of settlement (suburb, county, among others) and the name of the settlement. Thus, when the user selects a given city, the system automatically updates both its settlements and the corresponding establishments. After the parameters have been selected, results are displayed on a map by using

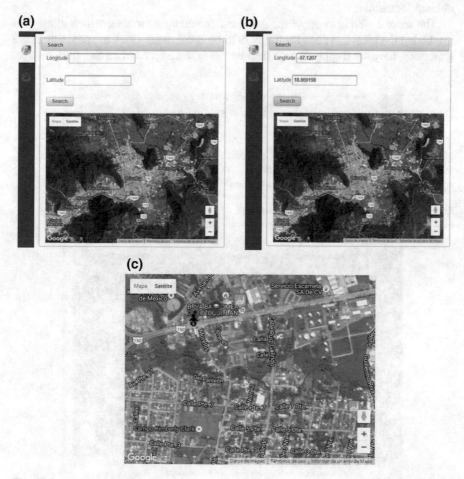

Fig. 12.3 Visualizing establishments for entering latitude and longitude by using GEOREMSYS. **a** Form before entering, **b** Form after entering, **c** Results

markers. The Fig. 12.2 shows results obtained from queries made to the Web
services of the APIs previously discussed in the case study section.

Figure (a) represents the visualization of the establishments's form before
selecting type and name of settlement and establishments category (schools,
nurseries, theaters, orphanages, retirement houses, hospitals, among others).
Figure (b) represents the visualization of the settlements's form after selecting.

The second functionality is the search of possible locations for POS by entering
latitude and longitude of the desired area as is shown in the Fig. 12.3a. First, the
user must enter the conditions for the establishment of POS. Such conditions or
restrictions determine the suitability of a location as illustrated in the Fig. 12.3b.

The Fig. 12.3c shows an entered location that does not meet the conditions and
is therefore inappropriate for a POS to be established on that area. As can be
observed, different signs indicate the existence of other establishments near the
entered location.

The second step is to apply the geo-recommendation process, which allows for
predicting the ideal locations for the POS to be established. Once the predictions are
generated, they are displayed on a map by using three types of circles (red, orange,

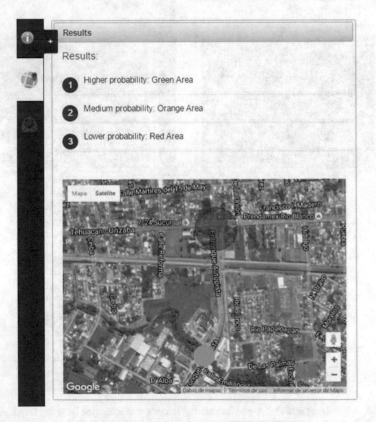

Fig. 12.4 Visualizing of geo-recommendations by using GEOREMSYS

and green). In other words, we can observe locations with high, medium, or low probability for the establishment of POS. The Fig. 12.4 shows results from the recommendation of different POS when the location meets the necessary conditions. As can be seen, green, orange, and red circles respectively indicate probable, mildly probable and less probable locations.

Circles appearing on the map help visualize in detail location results (probability, latitude, and longitude, and establishments settled around the area). The Fig. 12.5 shows results of a selected circle with the highest probability. Details show how appropriate the location is for settling a POS (probability percentage) and which establishments are found around the area (hospitals, retirement houses, nurseries, orphanages, schools, markets, churches, auditoriums, movie theaters, stadiums, and theaters).

Another functionality of GEOREMSYS is POS recommendation within a specific radial distance. Recommendation and search by radius allow for identifying suitable locations by entering a geographical point (longitude and latitude) and a

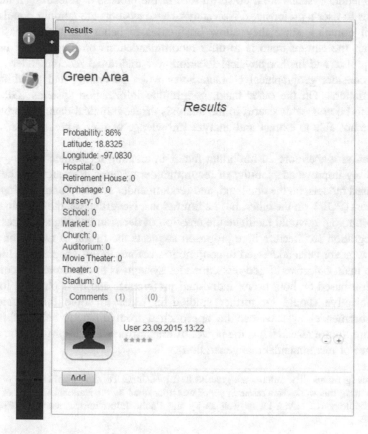

Fig. 12.5 Location details of a POS by using GEOREMSYS

radial distance. In this case, GEOREMSYS may find several locations for establishing a POS, and it shows an evaluation of each one of these locations. Also, within a same radius, the system can find more than one suitable location.

Recommendation by ZIP code allows GEOREMSYS to provide recommendations of POS within the same ZIP code. Here the system also performs an evaluation of all the possible locations. As in the previous functionality, the system may also detect more than one ideal location for establishing the POS. Finally, it is important to mention that GEOREMSYS cannot solve the cold-start problem, and this is its major limitation. However, it can offer geo-recommendations of POS locations depending on those previously established and the sales history.

12.5 Conclusions

Recommender systems are a powerful tool in the process of selecting an item that responds to user's preferences. Nowadays, these systems are gaining popularity in different domains, since they allow businesses to better know their customers. However, the current trend is to offer recommendations on the basis of both the user's profile and his/her physical location, yet traditional recommender systems fail to consider geographical characteristics when generating and providing recommendations. On the other hand, geographic information systems (GIS) have proven to be noticeably useful in the analysis of geographical data. Unfortunately, they are not able to extract and analyze knowledge of user's preferences on their own.

Location is therefore an important factor in creating recommendations, since it can visibly improve adaptability of recommender systems. Research on the use of GIS-based models, on the one hand, and recommender systems for analyzing points of interest (POIs), on the other, has its limitations. Nevertheless, the combination of both technologies would facilitate the creation of decision making strategies. Thus, the integration architecture here proposed suggests the fusion of two technologies that have never been addressed in conjunction before. In this chapter, we highlight that the main objective of geo-recommender systems is to consciously recommend a location based on both users' individual preferences and their physical location. This objective should be further studied in order to solve this problem, as geo-recommender systems can be approached from different fields of interest. Therefore, future research has the power of transforming and improving the current paradigm of recommendations generation.

Acknowledgements The authors are grateful to the National Technological Institute of Mexico for supporting this work. This research paper was sponsored by the National Council of Science and Technology (CONACYT), as well as by the Public Education Secretary (SEP) through PRODEP.

References

1. Colombo-Mendoza, L.O., et al.: RecomMetz: a context-aware knowledge-based mobile recommender system for movie showtimes. Expert. Syst. Appl. **42**(3), 1202–1222 (2015)
2. Noguera, J.M., et al.: A mobile 3D-GIS hybrid recommender system for tourism. Inf. Sci. **215**, 37–52 (2012)
3. http://www.retailangle.com/dictionary_result.asp?Keywords=P
4. García-Palomares, J.C., et. al.: Optimizing the location of stations in bike-sharing programs: a GIS approach. Appl. Geogr. **35**(1), 235–246 (2012)
5. Zolfani, S.H., et al.: Decision making on business issues with foresight perspective; an application of new hybrid MCDM model in shopping mall locating. Expert Syst. Appl. **40**(17), 7111–712 (2013)
6. Piastria, F.: Avoiding cannibalisation and/or competitor reaction in planar single facility location. J. Oper. Res. **48**(2), 148–157 (2005)
7. Batet, M., et al.: Turist@: agent-based personalised recommendation of tourist activities. Expert Syst. Appl. **39**(8), 7319–7329 (2012)
8. Li, Y.M., et al.: A social recommender mechanism for location-based group commerce. Inform. Sci. **274**, 125–142 (2014)
9. YU, Y.H., et al.: Recommendation system using location-based ontology on wireless internet: an example of collective intelligence by using 'mashup'applications. Expert Syst Appl. **36**(9), 11675–11681 (2009)
10. Pliakos, K., & Kotropoulos, C.: PLSA driven image annotation, classification, and tourism recommendation. In: 2014 IEEE International Conference on Image Processing (ICIP), IEEE (2014)
11. Yin, H., et al.: Trip mining and recommendation from geo-tagged photos. In: Multimedia and Expo Workshops (ICMEW), 2012 IEEE International Conference on, IEEE (2012)
12. Fu, C.Y., et al.: Travelbuddy: interactive travel route recommendation with a visual scene interface. In: International Conference on Multimedia Modeling. Springer International Publishing (2014)
13. Wei, L.Y., et al.: Discovering point-of-interest signatures based on group features from geo-social networking data. In: 2013 Conference on Technologies and Applications of Artificial Intelligence. IEEE (2013)
14. López-Ornelas, E, Abascal-Mena, R., Zepeda-Hernández, J.S.: A Geo-collaborative recommendation tool to help urban mobility. In: International Conference on Human-Computer Interaction. Springer International Publishing (2014)
15. Schedl, M., Schnitzer, D.: Location-aware music artist recommendation. In: International Conference on Multimedia Modeling. Springer International Publishing (2014)
16. Huang, L., Ma, Y., Liu, Y.: Point-of-Interest recommendation in location-based social networks with personalized geo-social influence. China Commun. **12**(12), 21–31 (2015)
17. Castro, C.A., et. al.: Modelación y simulación computacional usando sistemas de información geográfica con dinámica de sistemas aplicados a fenómenos epidemiológicos. Revista facultad de ingenieria Universidad de Antioquia. **34**, 86–100 (2005)
18. Corner, R.A., et al.: A fully integrated GIS-based model of particulate waste distribution from marine fish-cage sites. Aquaculture. **258**(1), 299–311 (2006)
19. Vairavamoorthy, K., et al.: IRA-WDS: a GIS-based risk analysis tool for water distribution systems. Environ. Model. Softw. **22**(7), 951–965 (2007)
20. Radiarta, I N , et al.: GIS based multi criteria evaluation models for identifying suitable sites for Japanese scallop (Mizuhopecten yessoensis) aquaculture in Funka Bay, southwestern Hokkaido, Japan. Aquaculture. **284**(1), 127–135 (2008)
21. Xu, Z, & Coors, V.: Combining system dynamics model, GIS and 3D visualization in sustainability assessment of urban residential development. Build. Environ. **47**, 272–287 (2012)

22. Suárez-Vega, R., et. al.: Location models and GIS tools for retail site location. Appl. Geogr. **35**(1), 12–22 (2012)
23. Roig-Tierno, N., et al.: The retail site location decision process using GIS and the analytical hierarchy process. Appl. Geogr. **40**, 191–198 (2013)
24. Casillas-Higuera, Á., et al.: Detección de la Isla Urbana de Calor mediante Modelado Dinámico en Mexicali, BC, México. Inf. Tecnol. **25**(1), 139–150 (2014)
25. Garg, D., Trivedi, K.: Fuzzy K-mean clustering in MapReduce on cloud based hadoop. In: Advanced Communication Control and Computing Technologies (ICACCCT), 2014 International Conference on. IEEE (2014)
26. Ingersoll, G.: Apache Mahout: scalable machine learning for everyone. IBM Corporat. (2011)
27. Gantner, Z., et al.: MyMediaLite: a free recommender system library. In: Proceedings of the Fifth ACM Conference on Recommender Systems, pp. 305–308. ACM (2011)
28. Gantner, Z., et al.: MyMediaLite: a free recommender system library. In: Proceedings of the Fifth ACM Conference on Recommender Systems. ACM (2011)
29. Caraciolo, M., Melo, B., & Caspirro, R.: Crab: a recommendation engine framework for python.
30. https://www.authorea.com/users/1713/articles/7010/_show_article
31. Hall, M., Frank, E., Holmes, G., Pfahringer, B., Reutemann, P., Witten, I.H.: The WEKA data mining software: an update. ACM SIGKDD Explor. Newsl. **11**(1), 10–18 (2009)
32. https://www.quora.com/How-can-we-use-Weka-tool-to-create-a-recommendation-system
33. Aksel, F., Birtürk, A.,: An adaptive hybrid recommender system that learns domain dynamics. In: International Workshop on Handling Concept Drift in Adaptive Information Systems: Importance, Challenges and Solutions (HaCDAIS-2010) at the European Conference on Machine Learning and Principles and Practice of Knowledge Discovery in Databases. (2010)
34. Aksel, F., Birtürk, A.: Enhancing Accuracy of Hybrid Recommender Systems through Adapting the Domain Trends. In: Workshop on the Practical Use of Recommender Systems, Algorithms and Technologies (PRSAT 2010). (2010)
35. https://developers.google.com/maps/documentation/geolocation/intro
36. Wen, M.: Exploring Spatial Analysis Capabilities in Google Maps Mashup Using Google Fusion Tables: A Case Study in Land Lease Data Retrieval. (2014)
37. Kanchev, G.M., Chopra, A.K.: Social media through the requirements lens: a case study of Google maps. In Crowd-Based Requirements Engineering (CrowdRE), 2015 IEEE 1st International Workshop on, pp. 7–12. IEEE (August 2015)
38. Fu, W., Nemesure, S.: Accelerator control data visualization with Google Map (No. BNL–100683-2013-CP). Brookhaven National Laboratory (BNL). (2013)
39. Tostes, A.I.J., et al.: From data to knowledge: city-wide traffic flows analysis and prediction using bing maps. In Proceedings of the 2nd ACM SIGKDD International Workshop on Urban Computing. ACM, pp. 12. (2013)
40. Penchalaiah, K., Sai Charan, V., LaNshmana Rao, B.: Efficient web based geospatial information system application using bing maps. Int. J. Adv. Res. Comput. Commun. Eng. **2**(10), 3916–3921 (2013)
41. Wagner, J.: Top 10 Mapping APIs: Google Maps, Microsoft Bing Maps and MapQuest (2015)
42. Eugster, M.J.A., Schlesinger, T.: osmar: OpenStreetMap and R. R J. **5**(1), 53–63 (2013):
43. H. Ayuntamiento de Córdoba, Ver. Reglamento Para Las Acciones De Construcción, Instalación, Conservación y Operación De Estaciones De Servicio En Gasolinera y Carburación para el Municipio de Córdoba, Ver, pp. 3–4 (2014)

Chapter 13
Evaluation of Denoising Methods in the Spatial Domain for Medical Ultrasound Imaging Applications

Humberto de Jesús Ochoa Domínguez and Vicente García Jiménez

Abstract Ultrasound is used as a real-time, non-invasive, portable, versatile and relatively low cost diagnostic imaging technique. The acquired images are corrupted by speckle noise that causes a low contrast in areas where lesion cannot be detected during the diagnosis stage. The characteristic of these images is that they follow a multiplicative noise model. Some techniques convert the multiplicative model into an additive model by transforming the image using the logarithm. Then, the noise is removed by denoising techniques designed for the additive model. In this chapter, the evaluation of denoising techniques, designed specifically for multiplicative noise models, applied in the spatial domain, is analyzed and compared using a synthetic image, a phantom image and real images. The aim of this study is to compare denoising methods when no transformation of the image is carried out.

13.1 Introduction

Ultrasound is a real-time, non-invasive, portable, versatile and relatively low cost, compared with other medical imaging techniques that allow the imaging of the internal body in real-time. Ultrasound has been widely accepted in the medical area since its emergence as a tool to help in the surgical guidance, preoperative planning and diagnose. This technique is preferred over techniques such as Magnetic Resonance Imaging (MRI) or the X-rays [1, 2], especially for the pregnant women, patients with arrhythmias or MR contraindications. For example, during a MRI study, the patient must remain very still to acquire a clear image and the powerful X-rays have the ability to create birth defects, diseases and can alter the DNA.

Ultrasound imaging technique uses ultrasonic waves produced from the transducer and travel through the patient's body. When the wavefront hits a disconti-

H.d.J. Ochoa Domínguez (✉) · V. García Jiménez
Departamento de Ingeniería Eléctrica y Computación,
Universidad Autónoma de Ciudad Juárez, Av. del Charro 450 norte,
Ciudad Juárez 32310, Chihuahua, México
e-mail: hochoa@uacj.mx

© Springer International Publishing AG 2017
G. Alor-Hernández and R. Valencia-García (eds.), *Current Trends on Knowledge-Based Systems*, Intelligent Systems Reference Library 120,
DOI 10.1007/978-3-319-51905-0_13

263

nuity, scattered waves are produced and echoes are bounced back. These echoes are detected by the same transducer, then processed and displayed by the equipment [3]. The noise is introduced by particles of tissues with size less than the wavelength applied and corrupts the ultrasound images. The interference of the sound waves scattered, received from different points in the imaged organ distorts important details during the acquisition stage. The resulting speckle noise pattern is visible in the image as dark and light spots of different sizes [4].

The speckle noise is one of the most important issues in ultrasound images because the poor signal to noise ratio may drastically affect the final diagnose. Therefore, the need of a denoising step may be vital to avoid errors in the diagnostic. The noise can be additive or multiplicative, depending on the modality used to acquire the image [5, 6]. Additive noise is easier to reduce and more tractable because the multiplicative noise is signal dependent.

The current methods to denoise make use of a trade-off between the reduction of unwanted data that distort the shape of the object of interest in the image and the image itself. The noise follows different probability distributions depending on the modality used. For example, in the ultrasound modality the model is multiplicative and follows a Rayleigh distribution. In other modalities, such as MRI and X-ray, the noise is Rician and Poisson distributed respectively [7].

In this chapter, an evaluation of denoising methods used to remove speckle noise, in the spatial domain, in ultrasound medical images, is carried out, with the aim of comparing their performance. The remaining of the chapter is organized as follows. In Sect. 13.2, a brief review of current methods in the state-of-the-art is presented and discussed. In Sect. 13.3, the multiplicative noise model of the ultrasound medical images is discussed. In Sect. 13.4, the filters used as test benches to remove speckle noise, against which all other methods compare their performance are discussed. In Sect. 13.5, the metrics used are explained. In Sect. 13.6, the phantom, the visual and the quantitative results of the filters shown in Sect. 13.5 are presented. Finally, the chapter concludes in Sect. 13.7.

13.2 Review of Speckle Noise Reduction Methods for Ultrasound Medical Images

The speckle noise can be reduced by increasing the ultrasound frequency. However, the higher the frequency contents the lower the penetration in the tissue [8, 9]. Consequently, most of the major commercial ultrasound systems software such as, Siemens, Kostec and GE ViewPoint 6, include Computer Aided Diagnosis (CAD) techniques for speckle noise reduction. CAD helps in the screening and avoids unnecessary biopsies [10–15]. Usually, texture features are used to discriminate normal, benign and malignant tissues in prostate analysis. Then, speckle filtering is an important pre-processing step in the analysis of ultrasound medical imaging. The speckle reduction methods can be classified into two main categories, those that process the image in the transform domain and those that process in the spatial domain (image). Even though in both categories the denoised image has good quality, the second category is preferred because is faster and can be used for real-time applications. However, extensive research has been done in both categories to yield good quality images in a reduced timeframe. Following, the most important works carried out in these two categories are presented.

13.2.1 Transform Domain Methods

Transform methods change the image to the frequency domain by using transforms such as, Fourier, wavelets, shearlet, curvelet and contourlet among others. After a filtering or a thresholding process, applied to the transform coefficients to reduce the noise, the denoised image is recovered by applying the inverse transform.

These methods have a high computational complexity due to the transformation and anti-transformation steps, and may insert artificial frequencies to the recovered image. Following, the most important methods found in the literature are explained.

In the Fast Fourier Transform (FFT) method, the image is enhanced, before transformation, to better detect the noise. Then, the FFT is applied to the image. The frequencies that represent dark and light spots (speckle) are searched, located and masked with a special function before recovering the image [16]. Notice that, besides the time introduced by the transformation, the searching operation is computationally expensive.

The wavelet transform is a multiscale and multi-resolution tool widely used in image processing [17]. Before transformation, the multiplicative noise model is transformed into an additive model by computing the logarithm of the image. Then thresholding is applied to shrink the coefficients. The main strength of wavelet thresholding is its capability to process the different frequency components separately [18–21]. Similarly, the Wiener filtering has been proposed as alternative to denoise the detail subbands [22]. Framelet algorithms, based on wavelet frames, combined with regularization terms, such as the total variation (TV) [23], have the advantages of multi-resolution analysis, remove noise and preserve edges [24, 25]. However, the use of TV may suppress the texture features of the image. The combination of wavelets, principal component analysis (PCA) and thresholding operations [26] has shown some improvements. Nevertheless, the calculation of the singular value decomposition of the covariance matrix, to compute the principal components, is computationally extensive. Furthermore, wavelets and non-linear diffusion [27] based on the iterative edge enhancement feature of nonlinear diffusion have shown interesting results in removing speckle noise while preserving the edges. Most of the efforts to despeckle ultrasound images have been done in the wavelet domain [28–35]. Nevertheless, wavelet-based methods require converting the image to the Cartesian space. This contributes to extra computational time and reduces solution accuracy.

Shearlet is multiscale transform, for signal representation, that provides directional selectivity, localization, anisotropy and shift invariance [36, 37]. Shearlet is an extension of wavelets and include the concept that the subbands capture all the anisotropic features such as edges. Wavelets cannot represent well the edges because they are isotropic objects [38]. Based on this, some works have been proposed to remove speckle noise in the shearlet domain. Most of them filter the transform coefficients to estimate the denoised shearlet coefficient. For example, an anisotropic diffusion method is applied to the noisy coefficients to reduce the noise and preserve the edge [39, 40].

Curvelets is another multiscale transform for representing very general linear symmetric systems of hyperbolic differential equations [41, 42]. Researchers have paid much attention to this transform [43–45]. The works estimate the unknown curvelet coefficients using similar filters as the used in the wavelet domain. For example, in [44], a threshold is calculated using SURE-LET [45] strategy and applied to the detail subbands. The results of shrinkage are further processed by using a nonlinear diffusion technique. In [46] the curvelet coefficients are modeled and processed with the Perona and Malik Anisotropic Diffusion filter [47]. A maximum a posteriori threshold is calculated to further process the coefficients and to avoid artifacts and to recover the edges.

The contourlet transform can deal effectively with smooth contours [48]. The pyramidal filter bank structure of the contourlet transform has very little redundancy. However, this transform is not shift-invariant. Conversely, the Non-Subsampled Contourlet Transform (NSCT) is fully shift-invariant, multi-scale and multi-direction expansion; most of the wavelet-based denoising methods have been extended to the NSCT for multiplicative noise removal assuming that the NSCT coefficients follow a generalized Gaussian distribution (GGD). However, this extension is not trivial [49–54].

13.2.2 Spatial Domain Methods

In the spatial domain denoising methods, no transformation is carried out on the original image. All the process is performed in the spatial domain (image domain). These methods are preferred for real-time application because they are less time-consuming [54]. Following, the most important methods found in the literature are explained.

In the non-local means (NLM) methods, the estimated pixel considers the weighted average of all pixels in the image. The weight depends on the similarity of the pixel under estimation with respect to another pixel usually inside a window. The weight depends on the noise deviation and the weighted Euclidean distance of the two pixels [55–57]. The algorithms are computationally expensive to be widely used in real-time applications. However, the recovered images show good removal of speckle noise and edge preservation.

The Oriented SRAD [58] is an extension of the Speckle Reducing Anisotropic Diffusion (SRAD) method [59] and is based on matrix anisotropic diffusion to obtain different diffusions across the principal curvature directions [60].

Methods based on the Total Variation (TV) [23] use a similarity term to measure the amount of smoothing in the image and a regularization term as a prior knowledge of the noise distribution. In ultrasound, the later term follows a Rayleigh distribution. The shortcoming is that noise samples can be large and an edge could be confused with a noise sample and vice versa. That is, if the noise is removed the edges can be blurred and if the edge is preserved, the noise could not be well-removed [61, 62].

The time-domain deconvolution methods yield good results. These methods rely on the Point Spread Function (PSF) calculation, for example the algorithms based on phase unwrapping and a noise-robust procedure to estimate the pulse in the complex cepstrum domain, then a robust estimate of the PSF can obtained to reduce the noise level linearly with the number of pulses estimates. The Wiener filter is used for subsequent deconvolution with sharper images than the original image because of the deconvolution. Other deconvolution method acts on the envelope of the acquired radio-frequency signals. The measured data is used to estimate a point spread function (PSF) the image is reconstructed in a non-blind way. These methods represent a major problem when the PSF has to be estimation [63–69].

13.3 Multiplicative Noise Model

The multiplicative noise, known as speckle, occurs in images acquired by coherent imaging systems such as Laser, SAR, Optical and ultrasound systems. This noise causes serious problems to represent an image. The intensity values of the pixels are multiplied by random values. The probability density function of the speckle noise follows a gamma distribution of the form:

$$P(x) = \frac{x^{\alpha-1}}{(\alpha-1)!a^{\alpha}} e^{-\frac{x}{a}}, \tag{3.1}$$

where x is the intensity of the pixel and the variance is $a^2\alpha$.

The observed samples are the output of the ultrasound imaging system and can be represented by the vector $\mathbf{g} = [g_1, \ldots, g_n]$. The samples of the speckle noise and the noise free image are the vectors $\mathbf{n} = [n_1, \ldots, n_n]$ and $\mathbf{f} = [f_1, \ldots, f_n]$ respectively. The additive noise vector is $\mathbf{\eta} = [\eta_1, \ldots, \eta_n]$. Therefore, the speckled image is commonly modeled as

$$\mathbf{g} = \mathbf{fn} + \mathbf{\eta}. \tag{3.2}$$

In general, it is widely accepted that the additive noise (from sensor) be considered small as compared to the multiplicative part. Hence, Eq. (3.2) can be reduced to:

$$\mathbf{g} = \mathbf{fn}. \tag{3.3}$$

Despeckling algorithms that use the model of Eq. (3.3) require two important conditions: preservation of the mean intensity value over homogeneous areas and preservation of important details such as texture and edges while removing the noise. The multiplicative speckle model was the base to development the minimum mean-square error (MMSE) Lee [70], Kuan et al. [71], Gamma MAP [72] and Frost et al. [73] filters. The Speckle Reducing Anisotropic Reduction (SRAD) [59] is based on the analysis of the Frost et al. and Lee filters as isotropic filters and the Perona and Malik model [47].

13.4 Speckle Denoising Filters Compared in This Chapter

Following, the comparison of filters that use the multiplicative model to remove speckle noise is described. We implemented the original version of every method As it can be seen in Sect. 13.2; many of the proposed methods could outperform the compared methods. However, they are hybrid methods (i.e. wavelets and thresholding or wavelets and TV or TV plus more terms), their performance depends on the input image and the selected transform and the implementation is computationally expensive. The following speckle filters are considered the test benches against which most of the methods are compared.

13.4.1 Average Filter

The image is processed using a sliding window of size $2k + 1$. It is also called neighborhood average method. It does not remove the speckle noise but integrates it incoherently into the mean value. In this procedure, the central pixel $\hat{f}$ of the sliding window is replaced by the average intensity, yielding a smooth image with blurred edges. The average filer is defined as:

$$\hat{f}_n = \frac{1}{2k+1} \sum_{i=-k}^{k} g_i. \tag{4.1}$$

The filter is optimal for images with additive Gaussian noise. In other words, it achieves the Kramer-Rao lower bound. However, it is not optimal for the case of a multiplicative model.

13.4.2 Median Filter

The median filter is a non-linear local procedure used to reduce speckle noise and to retain steps and ramp functions. The filter is robust to impulsive type noise with the particularity of preserving the edges of the image. Hence, it produces a less blurred image. The image is analyzed using a sliding window on the image in order to replace the central value of the window by its median intensity. In other words, the noisy pixel is replaced by its median value $\hat{f}_n$. Hence, the noise is reduced without blurring the edges. If the window length is $2k + 1$ the filtering is given by:

$$\hat{f}_n = med\,[g_{n-k}, \ldots, g_n, \ldots, g_{n+k}], \tag{4.2}$$

where $med[\cdot]$ is the median operator. To find the median it is necessary to sort all the intensities in a neighborhood into numerical order. This is a computationally complex process due to the time needed to sort pixels to find the median value of the window. Yet, the filter is good at preserving edges.

13.4.3 Frost Filter

The frost filter [73] is an adaptive Wiener filter that reduces the multiplicative noise while preserving edges. It replaces the central pixel of a window of size $2k + 1$ by the sum of weighted exponential terms. The weighting factors depend on the distance to the central pixel, the damping factor, and the local variance. The more far the pixel from the central pixel the less the weight. Also, the weighting factors increase as variance in the window increases. The filter convolves the pixel values within the window with the exponential impulse response:

$$h_i = e^{-K\alpha_g(i_0)|i|}, \tag{4.3}$$

where K is the filter parameter, i_0 is the location of the processed pixel, $|i|$ is the distance measured from the pixel i_0. The coefficient of variation is defined as $\alpha_g = \sigma_g/\bar{g}$; where $\bar{g}$ and σ_g are the local mean and standard deviation of the window. The optima MMSE Wiener filer in a window with normalization constant K_1 becomes:

$$\hat{f}_n = \sum_{i=-k}^{k} K_1 h_i g_i. \tag{4.4}$$

In order to preserve the mean value, K_1 is computed by using:

$$K_1 = \frac{1}{2k+1} \sum_{i=-k}^{k} h_i. \tag{4.5}$$

13.4.4 Kuan et al. Filter

The Kuan et al. filter [71] is a local minimum mean square error method used to restore images with signal-dependent noise. The filter considers the multiplicative noise model of Eq. (3.3) as an additive model of the form:

$$\mathbf{g} = \mathbf{f} + (\mathbf{n} - 1)\mathbf{f}. \tag{4.6}$$

Assuming unit-mean noise, the estimate pixel value $\hat{f}_n$ in the local window is:

$$\hat{f}_n = \bar{g} + \frac{\sigma_f^2 (g_n - \bar{g})}{\sigma_f^2 (ENL + 1)\bar{g}^2} ENL, \tag{4.7}$$

with

$$\sigma_f^2 = \frac{ENL\sigma_g^2 - \bar{g}^2}{ENL + 1}, \tag{4.8}$$

and

$$ENL = \left(\frac{Mean}{StDev}\right)^2 = \left(\frac{\bar{g}}{\sigma_g^2}\right)^2. \tag{4.9}$$

The Equivalent Number of Looks (ENL) estimates the noise level and is calculated in a uniform region of the image. One shortcoming of this filter is that the ENL parameter needs to be computed beforehand.

13.4.5 Lee Filter

The Lee filter [70] uses the Digital Number (DN) values calculated in the window to estimate the pixel under processing. Unlike a typical low-pass smoothing filter, the Lee filter and other similar sigma filters preserve image sharpness and details while reducing the noise. The pixel being filtered is replaced by a value that is calculated using the surrounding pixels. If the variance is low, smoothing will be performed. In the other hand, if the variance is high, assuming an edge, the smoothing will not be performed. Therefore, Eq. (4.7) can be simplified as:

$$\hat{f}_n = \bar{g} + k(g_n - \bar{g}), \tag{4.10}$$

where k is an adaptive filter coefficient. The Lee filter is a particular case of the Kuan et al. filter without the term σ_f^2/ENL.

13.4.6 Gamma MAP Filter

The Gamma Maximum A Posteriori (MAP) filter [72] is based on a Bayesian analysis of a multiplicative noise model and assumes that the image and the noise follow a Gamma distribution. The pixel being filtered is replaced with a value that is calculated based on the local statistics. The filter takes into consideration not only the noise but also the image. The Gamma MAP filter of the pixel under processing is given by:

$$\hat{f}_n = \frac{\bar{g}(\alpha - ENL - 1) + \sqrt{4\alpha\, ENL\, g\, \bar{g} + \bar{g}^2(\alpha - ENL - 1)^2}}{2\alpha}, \tag{4.11}$$

where

$$\alpha = \frac{ENL + 1}{ENL(\sigma_g/\bar{g})^2 - 1}. \tag{4.12}$$

13.4.7 Anisotropic Diffusion

Diffusion is a physical phenomenon that is used in the image-processing field to denoise images and to detect edges. The phenomenon aims at minimizing the spatial concentration $g(x; t)$ of a substance. In other words, the goal is to minimize the differences in values of the pixels belonging to similar regions. This process is described by the Fick's law that states that concentration differences induce a flow j of a substance in direction of the negative concentration gradient. Therefore, the flow can be expressed as:

$$j = -c\nabla g, \tag{4.13}$$

where c is the diffusivity that describes the speed of the diffusion process from one polnt to another and ∇ (nabla) is the gradient operator. Once the flow is described, the continuity equation is used to observe the change in time of the concentration g and is expressed by the negative of the divergent of the flow:

$$\partial_t g = -div\, j. \qquad (4.14)$$

The diffusion equation is obtained by replacing Eq. (4.13) into Eq. (4.14) as:

$$\partial_t g = div(c\nabla g). \qquad (4.15)$$

If the diffusivity c is a constant, i.e. $c = 1$, the process is linear isotropic and homogeneous. If c depends on the concentration g, $c = c(g)$, and the process becomes a nonlinear diffusion. However, if c is a matrix-valued diffusivity, the process is called anisotropic and it will lead to a process where the diffusion is different for different directions. In image processing, the goal is to use the anisotropic process to obtain less diffusion in edges. In other words, the diffusivity should decrease with strong gradients. Perona and Malik [47] proposed a generalization of the diffusion equation to denoise images without blurring the edges. Base on this, the speckle reducing anisotropic diffusion (SRAD) was later proposed by Yongjian and Acton [59]. The approach is based on the minimum mean square error (MMSE) approach of the Lee and Frost filters and extended the Perona and Malik algorithm for images corrupted by speckle noise.

13.4.7.1 Speckle Reducing Anisotropic Diffusion (SRAD)

Yongjian and Acton [59] rearranged the Eq. (4.7) as:

$$\hat{f} = g + (1 - k)(\bar{g} - g). \qquad (4.16)$$

The term $(\bar{g} - g)$ can be seen as an approximation to the Laplacian operator (with $c = 1$). Then, Eq. (4.16) can be expressed as:

$$\hat{f} = g + k'\, div\, (\nabla g). \qquad (4.17)$$

Equation (4.17) is an isotropic process. Hence, Eq. (4.15) can be easily transformed into an anisotropic version by including only the c factor:

$$\partial_t g = div(c\nabla g) = c\, div\, (\nabla g) + \nabla c \nabla g. \qquad (4.18)$$

The SRAD filter can be stated as follows. Given an initial condition that indicates the amount of concentration at time $t = 0$, with finite energy and no zero values over the image domain Ω. The output image $g(x, y; t)$ is evolved according to the following Partial Derivative Equation (PDE).

$$\begin{cases} \frac{\partial g(x,y;t)}{\partial t} = div\left[c(q)\nabla g(x,y;t)\right] \\ g(x,y;0) = g(x,y) \\ \left.\frac{\partial g(x,y;t)}{\partial \bar{n}}\right|_{\partial \Omega} = 0 \end{cases} \tag{4.19}$$

Observe that $c(q)$ is the diffusion coefficient and $g(x, y; t)$ is the instantaneous coefficient of variation. In other words, it is the edge detector. The last boundary condition states that, the derivative of the function along the outer normal, at the image boundary, must vanish. This assures that no concentration (brightness) will leave or enter the image, i.e. the average brightness will be preserved.

13.5 Metrics

Besides the mean and variance of the reconstructed images, the filters were evaluated on the synthetic image using the mean square error (*MSE*) and in the cyst phantom using the contrast to noise ratio (*CNR*) and the lesion to background contrast (*CLB*) [74]. The metrics are defined as follows:

$$MSE = \frac{1}{M \times N} \sum_{m=0}^{M-1} \sum_{n=0}^{N-1} \left[I(m,n) - \hat{I}(m,n)\right]^2. \tag{5.1}$$

$M \times N$ is the image size, I is the clean or reference image and $\hat{I}$ is the filtered image.

$$CNR = \frac{|\mu_L - \mu_B|}{\sqrt{\sigma_B^2 + \sigma_L^2}}. \tag{5.2}$$

$$CLB = \frac{\mu_L - \mu_B}{\mu_B}. \tag{5.3}$$

Notice that μ_L and σ_L^2 are the mean and variance in the lesion (cyst) respectively. μ_L and σ_B^2 are the mean and variance of intensities of pixels in the background region.

13.6 Results

The synthetic datum is a 256×256 image, gray scale, 8 bits per pixel. The image was contaminated with speckle noise of variance $\sigma_N = 0.02$ and $\sigma_N = 0.02$. The variance, mean and *MSE* of the denoised image was calculated with respect to the clean image (true image). To evaluate the *CNR* and the *CLB* a cyst from a B-mode image cyst Phantom-Field II ultrasound simulation [75, 76] was used. The experiments were carried out in an Intel core i5 1.60 GHz processor with the visual C# software.

Figures 13.1 and 13.2 show the resulting images for two different noise variances. Table 13.1 shows the quantitative results. We can observe that the Frost, the Lee, the Kuan, the Gamma MAP and the SRAD filters yield sharper recovered images sharper images and in the case of the Median and the Frost filters most of the edges are preserved but the images are blurred.

In Table 13.1 we can see that the Median, Frost et al., Lee, Kuan et al., Gamma MAP and SRAD filters increased the intensity value up to a certain point but the mean is not well preserved. The Average filter decreased the mean and produced a more blurred image. One reason for this is the multiplicative nature of speckle noise, which relates the amount of noise to the signal intensity. The other reason is that the filter is not adaptive in the sense that do not account for the

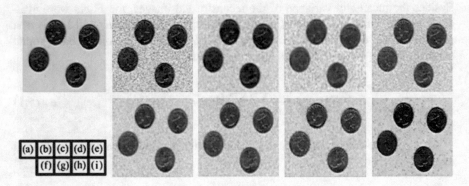

Fig. 13.1 Synthetic image used to compare the algorithms. **a** Clean image. **b** Noisy image $\sigma_N = 0.02$. Images filtered by using the **c** mean, **d** median, **e** Frost et al., **f** Lee, **g** Kuan et al., **h** Gamma MAP and **i** SRAD filters

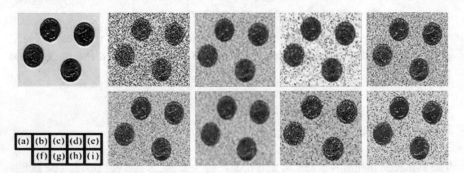

Fig. 13.2 Synthetic image used to compare the algorithms. **a** Clean image. **b** Noisy image. $\sigma_N = 0.2$ Images filtered by using the **c** mean, **d** median, **e** Frost et al., **f** Lee, **g** Kuan et al., **h** Gamma MAP and **i** SRAD filtered images

Table 13.1 Mean, standard deviation and MSE of the recovered synthetic images after filtering

Filter	$\sigma_N = 0.02$			$\sigma_N = 0.2$		
	Mean	St. Dev.	MSE	Mean	St. Dev.	MSE
Noisy image (no filter)	0.7667	0.2238	638.27	0.6855	0.2938	4657.3
Mean	0.7571	0.2044	153.33	0.6780	0.1558	974.7
Median	0.7412	0.2192	206.84	0.6948	0.2314	763.7
Frost et al.	0.7683	0.2041	233.87	0.7240	0.2014	1766.0
Lee	0.7665	0.2186	146.31	0.7053	0.1806	1089.1
Kuan et al.	0.7962	0.2055	126.97	0.7011	0.1780	839.0
Gamma MAP	0.7595	0.2138	118.51	0.7196	0.1939	1450.2
SRAD-500 it	0.7892	0.2456	302.71	0.7108	0.2139	870.0

Mean and standard deviation of the clean image is 0.7775 and 0.2072 respectively

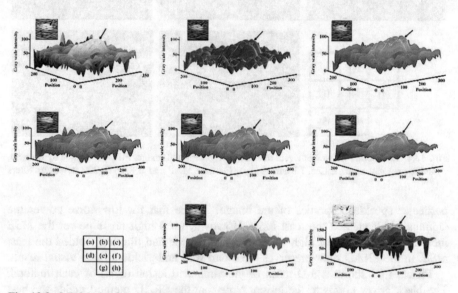

Fig. 13.3 3-D Mesh of the simulated B-mode cyst Phantom Field-II used to compute the *CNR* and *CLB*. **a** Original image. Mesh after processing with the **b** mean, **c** median, **d** Frost et al., **e** Lee, **f** Kuan et al., **g** Gamma MAP and **h** SRAD-200 iterations filters

Table 13.2 Results of Contrast to Noise Ratio (CNR) and Lesion to Background Contrast (CLB) fort the simulated B-mode cyst Phantom Field-II

Filter	CRN	CLB
Original cyst image (no filter)	0.5222	0.5819
Mean	0.3984	0.5828
Median	0.4213	0.5773
Frost et al.	0.8373	0.5820
Lee	0.7048	0.5831
Kuan et al.	0.9466	0.5838
Gamma MAP	0.5222	0.5836
SRAD-200 iterations	0.3984	0.5851

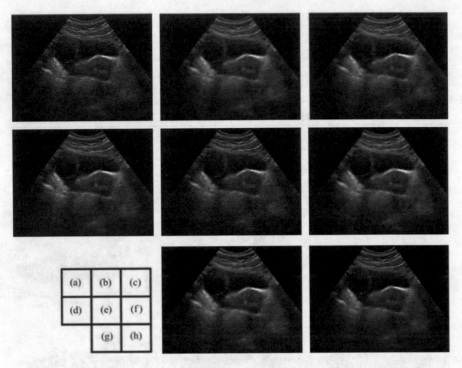

Fig. 13.4 Ultrasound images of a real ovarian cyst. **a** Original image. Filtered images by **b** mean, **c** median, **d** Frost et al., **e** Lee, **f** Kuan et al., **g** Gamma MAP and **h** SRAD–200 iterations filters

particular speckle properties of the image. Notice that for low noise power the Gamma MAP yielded the least *MSE*. However, for a high noise power the *MSE* increased. In the case of high variance noise the Median filtering yielded the least *MSE*. In the SRAD filter preserves the mean well and yields a good visual result.

Figure 13.3 shows a 3-D mesh of the simulated lesion (cyst) of each method. The black arrow points to the lesion. Note that the SRAD method yields the best result because it reduces the noise and preserves the lesion yielding the best *CLB* as shown in Table 13.2. However, SRAD took 200 iterations to obtain the final result. Also, the Kuan et al., and Gamma MAP filters yield better results than the remaining of the filters.

The quantitative results for the cyst phantom are shown in Table 13.2. It can be seen that the SRAD yields the best result in 200 iterations.

The results of filtering a real ultrasound image with an ovarian cyst are shown in Fig. 13.4. It can be observed how SRAD filter performs well however the image texture is lost.

13.7 Conclusions

Ultrasound is becoming more useful for medical diagnostics. Speckle introduced during the image acquisition distorts the image and can lead to erroneous diagnostics. Medical application software's include post-processing techniques to reduce the speckle that contaminates the ultrasound images and improve the shape of the images to help in the diagnostic. In this chapter, several types of filters, applied in the spatial domain of the image, have been evaluated. It was seen that the Median filter can preserve the edges; however, the mean of the image is altered. The Lee, the Frost, and the Gamma Map filter seem similar in reducing speckle noise. Nevertheless, Lee, Kuan and Frost yield better results in reducing speckle in texture areas. Gamma MAP preserves more texture while SRAD removes more noise and texture. The denoising methods discussed have been integrated in a medical tool, developed in visual C# to enhance images and help in the analysis of medical image in the patient's diagnostic stage.

Acknowledgements The authors express their gratitude to the Ultrasound Specialist, Dr. Moira Cuervo-Trigo for the valuable discussions, for testing the software and for providing the ultrasound images.

References

1. Wiell, C., Szkudlarek, M, Hasselquist, M., Møller, J. M., Vestergaard, A., Nørregaard, J., Terslev, L., Østergaard, M.: Ultrasonography, magnetic resonance imaging, radiography, and clinical assessment of inflammatory and destructive changes in fingers and toes of patients with psoriatic arthritis. J. Arthritis Res. Ther. 1–13 (2007)
2. Wright, G.: Magnetic resonance imaging. IEEE Sig. Process. Mag. 56–66 (1997)
3. Erikson, K.R., Fry, F.J., Jones, J.P.: Ultrasound in medicine-a review. IEEE Trans. Sonics Ultrason. **21**(3), 144–170 (1974)
4. Tay, P.C., Acton, S.T., Hossack. J.A.: A stochastic approach to ultrasound despeckling. In 3rd IEEE International Symposium on Biomedical Imaging: Nano to Macro, Arlington, VA (2006)
5. Gonzalez, R.C., Woods, R.E.: Digital Image Processing, 3rd edn. Pearson P.H, Saddle River, NJ (2007)
6. Chang, Q., Yang, T.: A Lattice Boltzmann method for image denoising. IEEE Trans. Image Process. **18**(12), 2797–2802 (2009)
7. Nowak, R.D.: Wavelet-based Rician noise removal for magnetic resonance imaging. IEEE Trans. Image Process. 1408–1419 (1999)
8. Goodman, J.W.: Some fundamental properties of speckle. J. Opt. Soc. America **66**(1), 1145–1150 (1976)
9. Ragesh, N.K, Reghunadhan, R, Anil, A.R.: Digital image denoising in medical ultrasound images: a survey. In International Conference on Artificial Intelligence and Machine Learning, AIML-11, Dubai, United Arab Emirates (2011)
10. Hadjerci, O., Hafiane, A., Conte, D., Makris, P., Vieyres, P., Delbos, A.: Computer-aided detection system for nerve identification using ultrasound images: a comparative study. Inform Med Unlocked **3**, 29–43 (2016)

11. Hadjerci, O., Hafiane, A., Morette, N., Novales, C., Vieyres, P., Delbos, A.: Assistive system based on nerve detection and needle navigation in ultrasound images for regional anesthesia. Expert Syst. Appl. **61**(1), 64–77 (2016)
12. Maggio, S., Palladini, A., Marchi, L.D., Alessandrini, M., Speciale, N., Masetti, G.: Predictive Deconvolution and hybrid feature selection for computer-aided detection of prostate cancer. IEEE Trans. Med. Imaging **29**(2), 455–464 (2010)
13. Frenkel, O., Mansour, K., Fischer, J.W.: Ultrasound-guided femoral nerve block for pain control in an infant with a femur fracture due to non-accidental trauma. Pediatr. Emerg. Care **28**(2), 183–184 (2012)
14. Bernardes, M.C., Adorno, B.V., Poignet, P., Borges, G.A.: Robot-assisted automatic insertion of steerable needles with closed-loop imaging feedback and intraoperative trajectory replanning. Mechatronics **23**(6), 630–645 (2013)
15. Yanong, Z., Stuart, W., Reyer, Z.: Computer technology in detection and staging of prostate carcinoma: a review. Med. Image Anal. **10**(2), 178–199 (2006)
16. Liu, J.G., Keith-Morgan, G.L.: FFT selective and adaptive filtering for removal of systematic noise in ETM+ imageodesy images. IEEE Trans. Geosci. Remote Sens. **44**(12), 3716–3724 (2006)
17. Pizurica, A., Philips, W., Lemahieu, I., Acheroy, M.: A versatile wavelet domain noise filtration technique for medical imaging. IEEE Trans. Med. Imaging **22**(3), 323–331 (2003)
18. Karthikeyan, K., Chandrasek, C.: Speckle noise reduction of medical ultrasound images using Bayesshrink Wavelet Threshold. Int. J. Comput. Appl. **22**(9), 8–14 (2011)
19. Vishwa, A., Sharma, S.: Speckle noise reduction in ultrasound images by Wavelet Thresholding. Int. J. Adv. Res. Comput. Sci. Softw. Eng. **2**(1), 525–530 (2012)
20. Sudha, S., Suresh, G.R., Sukanesh, R.: Speckle noise reduction in ultrasound images by Wavelet Thresholding based on weighted variance. Int. J. Adv. Res. Computer Sci. Softw. Eng. **1**(1), 7–12 (2009)
21. Dhillon, H., Jindal, G.D., Girdhar, A.: A novel Threshold technique for eliminating speckle noise in ultrasound images. In: International Conference on Modeling, Simulation and Control, Singapore (2011)
22. Udomhunsakul, S., Wongsita, P.: Ultrasonic speckle denoising using the combination of wavelet transform and wiener filter. In: Proceedings of the International Conference on Computing Intelligence and Computing Research, Bangkok (2004)
23. Rudin, L.I., Osher, S., Fatemi, E.: Nonlinear total variation based noise removal algorithms. Physica D **60**(1), 259–268 (1992)
24. Cai, J.-F., Dong, B., Osher, S., Shen, Z.: Image restoration: total variation, wavelet frames, and beyond. J. Amer. Math. Soc **25**(1), 1033–1089 (2012)
25. Abrahim, B.A., Kadah, Y.: Speckle noise reduction method combining total variation and wavelet shrinkage for clinical ultrasound imaging. In: 1st Middle East Conference on Biomedical Engineering, Cairo (2011)
26. Jagadesh, T., Rani, R.J.: A novel speckle noise reduction in biomedical images using PCA and wavelet transform. In: 2016 International Conference on Wireless Communications, Signal Processing and Networking (WiSPNET), Chennai (2016)
27. Yong, Y., Croitoru, M.M., Bidani, A., Zwischenberger, J.B., Clark, J.W.: Nonlinear multiscale wavelet diffusion for speckle suppression and edge enhancement in ultrasound images. IEEE Trans. Med. Imaging **25**(3), 297–311 (2006)
28. Zhang, X., Zhang, S.: Diffusion scheme using mean filter and wavelet coefficient magnitude for image denoising. AEU—Int. J. Electron. Commun. **70**(7), 944–952 (2016)
29. Kishore, P.V.V., Sastry, A.S.C.S., Kartheek, A., Mahatha, S.H.: Block based thresholding in wavelet domain for denoising ultrasound medical images. In Signal Processing And Communication Engineering Systems (SPACES), 2015 International Conference on, Vijayawada (2015)
30. Khare, A., Khare, M., Jeong, Y., Kim, H., Jeon, M.: Despeckling of medical ultrasound images using Daubechies complex wavelet transform. Sig. Process. **90**(2), 428–439 (2010)

31. Thakur, A., Anand, R.S.: Image quality based comparative evaluation of wavelet filters in ultrasound speckle reduction. Digit. Sig. Proc. **15**(5), 455–465 (2005)
32. Esakkirajan, S., Vimalraj, C.T., Muhammed, R., Subramanian, G.: Adaptive Wavelet packet-based de-speckling of ultrasound images with bilateral filter. Ultrasound Med. Biol. **39** (12), 2463–2476 (2013)
33. Gupta, S., Chauhan, R.C., Sexana, S.C.: Wavelet-based statistical approach for speckle reduction in medical ultrasound images. Med. Biol. Eng. Compu. **42**(2), 189–192 (2004)
34. Zhang, J., Guangkuo, L., Wu, L., Wang, C., Yun, C.: Wavelet and fast bilateral filter based de-speckling method for medical ultrasound images. Biomed. Sig. Process. Control **18**(1), 1–10 (2015)
35. Sudarshan, V.K., Mookiah, M.R.K., Acharya, U.R., Chandran, V., Molinari, F., Fujita, H., Ng, K.H.: Application of wavelet techniques for cancer diagnosis using ultrasound images: a Review. Comput. Biol. Med. **69**(1), 97–111 (2016)
36. Guo, K., Kutyniok, G., Labate, D.: Sparse multidimensional representations using anisotropic dilation and shear operators. In: Wavelets and Splines, pp. 189–201, Nashville, TN, Nashboro Press (2006)
37. Labate, D., Lim, W.-Q., Kutyniok, G., Weiss, G.: Sparse Multidimensional Representation Using Shearlets. In: SPIE 5914, Wavelets XI, San Diego, CA (2005)
38. Mallat, S.: A Wavelet Tour of Signal Processing: The Sparse Way, Burlington. Academic Press, MA (2008)
39. Deep, G., Anand, R.S., Barjeev, T.: Despeckling of ultrasound medical images using nonlinear adaptive anisotropic diffusion in nonsubsampled shearlet domain. Biomed. Sig. Process. Control **14**(1), 55–65 (2014)
40. Deep, G., Anand, R.S., Barjeev, T.: Speckle filtering of ultrasound images using a modified non-linear diffusion model in non-subsampled shearlet domain. IET Image Proc. **9**(2), 107–117 (2015)
41. Candès, E.J., Donoho, D.L.: Curvelets and curvilinear integrals. J. Approximation Theor. **113** (1), 59–90 (2001)
42. Starck, J.-L., Candès, E.J., Donoho, D.L.: The curvelet transform for image denoising. IEEE Trans. Image Process. **11**(6), 670–684 (2002)
43. Devarapu, K.V., Murala, S., Kumar, V.: Denoising of ultrasound images using curvelet transform. In: 2010 The 2nd International Conference on Computer and Automation Engineering (ICCAE), Singapore (2010)
44. Binjin, C., Yang, X., Jianguo, Y.: Ultrasonic speckle suppression based on a novel multiscale thresholding technique. In: 5th International Symposium on I/V Communications and Mobile Network (ISVC), Rabat (2010)
45. Stein, C.M.: Estimation of the mean of a multivariate normal distribution. Ann. Statist. **9**(1), 1135–1151 (1981)
46. Bama, S., Selvathi, D.: Despeckling of medical ultrasound kidney images in the curvelet domain using diffusion filtering and MAP estimation. Sig. Process. **103**(1), 230–241 (2014)
47. Perona, P., Malik, J.: Scale-space and edge detection using anisotropic diffusion. IEEE Trans. Pattern Anal. Mach. Intell. **12**(7): 629–639 (1990)
48. Do, M.N., Vetterli, M.: The contourlet transform: an efficient directional multiresolution image representation. IEEE Trans. Image Process. **14**(12), 2091–2106 (2005)
49. Song, X.-Y., Chen, Y.-Z., Zhang, S., Yang, W.: Speckle reduction based on contourlet transform using scale adaptive threshold for medical ultrasound image. J. Shanghai Jiaotong University (Science) **13**(5), 553–558 (2008)
50. Hiremath, P.S., Akkasaliga, P.T., Badige, S.: Speckle reducing contourlet transform for medical ultrasound images. Int. J. Comput. Electr. Autom, Control Inf. Eng. **4**(4), 284–291 (2011)
51. Xuhui, C., Lei, L., Hui, L., Peirui, B.: Ultrasound image denoising based on the contourlet transform and anisotropic diffusion. In Seventh International Conference on Image and Graphics (ICIG), Qingdao, Shandong (2013)

52. Argenti, F., Alparone, L.: Speckle removal from SAR images in the undecimated wavelet domain. IEEE Trans. Geosci. Remote Sens. **40**(11), 2363–2374 (2002)
53. Abd-Elmoniem, K.Z., Youssef, A.B., Kadah, Y.M.: Real-time speckle reduction and coherence enhancement in ultrasound imaging via nonlinear anisotropic diffusion. IEEE Trans. Biomed. Eng. **49**(9), 997–1014 (2002)
54. Xin, Z., Xili, J.: Image denoising in contourlet domain based on a normal inverse Gaussian prior. Digit. Sig. Proc. **20**(2), 1439–1446 (2010)
55. Coupe, P., Hellier, P., Kervrann, C.: Nonlocal means-based speckle filtering for ultrasound images. IEEE Trans. Image Process. **18**(10), 2221–2229 (2009)
56. Guo, Y., Wang. Y., Hou, T.: Speckle filtering of ultrasonic images using a modified non local-based algorithm. Biomed. Sig. Process. Control. **6**(2), 129–138 (2011)
57. Sudeep, P.V., Palanisamy, P., Rajan, J., Baradaran, H., Saba, L., Gupta, A., Suri, J.S.: Speckle reduction in medical ultrasound images using an unbiased non-local means method. Biomed. Sig. Process. Control. **28**(1): 1–8 (2016)
58. Krissian, K., Westin, C.-F., Kikinis, R., Vosburgh, K.G.: Oriented speckle reducing anisotropic diffusion. IEEE Trans. Image Process. **16**(5), 1412–1424 (2007)
59. Yu, Y., Acton, S.T.: Speckle reducing anisotropic diffusion. IEEE Trans. Image Process. **11** (11), 1260–1270 (2002)
60. Krissian, K., Kikinis, R., Vosburgh, K.: Speckle-constrained filtering of ultrasound images. In IEEE Computer Society Conference on Computer Vision and Pattern Recognition (CVPR'05), San Diego, CA (2005)
61. Denis, L., Tupin, F., Darbon, J., Sigelle, M.: SAR image regularization with fast approximate discrete minimization. IEEE Trans. Image Process. **18**(7), 1588–1600 (2009)
62. Hacini, M., Hachouf, F., Djemal, K.: A new speckle filtering method for ultrasound images based on a weighted multiplicative total variation. Sig. Process. **103**(1), 214–229 (2014)
63. Chira, L.-T., Rusu, C., Girault, J.-M.: Speckle noise removal in ultrasound medical imaging using envelope based time domain deconvolution. In International Symposium on Signals, Circuits and Systems (ISSCS), Iasi, Romania (2013)
64. Sabo, T.L.: Diagnostic Ultrasound Imaging: Inside Out, 2nd edn. Academic Press, San Diego, CA (2013)
65. Bishop, T., Badacan, S., Amizic, B., Katsaggelos, A., Chan, T., Molina, R.: Classification of Blind Image Deconvolution Methodologies. In: Blind Image Deconvolution: Theory and Applications, Boca Raton, CRC Press (2007)
66. Hundt, E., Trautenberg, E.: Digital processing of ultrasonic data by deconvolution. IEEE Trans. Sonics Ultrason. **27**(5), 249–252 (1980)
67. Abeyratne, U., Petropulo, A., Reid, J.: Higher order spectra based deconvolution of ultrasound images. IEEE Trans. Ultrason. Ferroelectr. Freq. Control **42**(6), 1064–1075 (1995)
68. Jirik, R., Taxt, T.: Two-dimensional blind Bayesian deconvolution of medical ultrasound images. IEEE Trans. Ultrason. Ferroelectr. Freq. Control **55**(10), 2140–2153 (2008)
69. Michailovich, O., Tannenbaum, A.: Blind deconvolution of medical ultrasound images: a parametric inverse filtering approach. IEEE Trans. Image Process. **16**(12), 3005–3019 (2007)
70. Jong-Sen, L.: Digital image enhancement and noise filtering by use of local statistics. IEEE Trans. Pattern Ana. Machine Intell. **PAMI-2**(2): 165–168 (1980)
71. Kuan, D.T., Sawchuk, A.A., Strand, T.C., Chavel, P.: Adaptive noise smoothing filter for images with signal-dependent noise. IEEE Trans. Pattern Ana. Mach. Intell. **PAMI-7**(2), 165–177 (1985)
72. Lopes, A., Nezry, E., Touzi, R., Laur, H.: Maximum a posteriori speckle filtering and first order texture models in SAR images. In: IEEE International Geoscience and Remote Sensing Symposium (IGARSS), College Park, MD (1990)
73. Frost, V.S., Stiles, J.A., Shanmugan, K.S., Holtzman, J.C.: A model for Radar Images and Its Application to Adaptive Digital Filtering of Multiplicative Noise. IEEE Transactions on Pattern Analysis and Machine Intelligence **PAMI-4**(2), 157–166 (1982)

74. Chan, C., Fulton, R., Feng, D.D., Meikle, S.: Median non-local means filtering for low SNR image denoising: application to PET with anatomical knowledge. In: IEEE Nuclear Science Symposium & Medical Imaging Conference, Knoxville, TN (2010)
75. Jensen, J.A., Svendsen, N.B.: Calculation of pressure fields from arbitrarily shaped apodized and exited ultrasound transdurer. IEEE Trans. Ultrason. Ferroelectr. Freq. Control **39**(1), 262–267 (1992)
76. Jensen, J.A.: Field II simulation program, 1 Jan 2013 [Online]. Available: http://field-ii.dk. Accessed 3 Mar 2015

Appendix A
Attributes for AMT Ergonomic Evaluation

© Springer International Publishing AG 2017
G. Alor-Hernández and R. Valencia-García (eds.), *Current Trends on Knowledge-Based Systems*, Intelligent Systems Reference Library 120, DOI 10.1007/978-3-319-51905-0

		Karwowski [1], Chaps. 8, 10, 12, 16, 19 and 21	Lee [2]	Genaidy et al. [3]	Genaidy et al. [4]	Bruseberg [5]	Ruckart and Burgess [6]	Siemieniuch and Sinclair [7]	Stanton [8]	Kesseler and Knapen [9]	Besnard and Cacitti [10]	Vieira and Kumar [11]	Balogh et al. [12]
Ergonomic compatibility-main attributes (ECMA)	A11		A131, A132, A133, A134, A135, A137			A111, A112			A112				
	A12	A122, A123, A124, A125		A124, A125	A121, A122, A123, A124, A125	A121, A122, A123, A124, A125,						A124, A125	A124, A125
	A13	A131, A132, A133, A134, A135, A137				A131, A132, A133, A134, A135, A136, A137	A134, A135, A136	A134, A135, A136	A134, A135, A136	A133, A134, A135	A133, A134, A135		
	A14	A141, A142, A143			A141, A142, A143								
	A15			A151, A152	A151, A152				A151, A152			A151, A152	

(continued)

(continued)

Source		Maldonado-Macias et al. [13]	Maldonado-Macias et al. [13]	Chan et al. [14]	Tavana et al. [15]	Chuu [16]	Chuu [17]	Ramnath et al. [18]	Neumann and Medbo [19]	Battini et al. [20]	Jayaram et al. [21]	Rossi et al. [22]	Battiniet et al. [23]
Ergonomic compatibility-main attributes (ECMA)	A11	A111, A112		A111, A112	A112			A111, A112				A112	
	A12	A121, A122, A123, A124, A125			A121	A121, A122	A121, A122	A121, A122, A123, A124, A125	A121, A122, A123, A124, A125	A121, A122, A123, A124, A125	A121, A122, A123, A124, A125	A121, A122, A123, A124, A125	A121, A122, A123, A124, A125
	A13	A131, A132, A133, A134, A135, A136 A137	A131, A132, A133, A134, A135, A136 A137	A131, A132, A133, A134, A135, A136 A137	A137			A134		A136		A131, A132, A133, A134, A135, A136, A137	
	A14	A141, A142 A143, A144			A144							A141, A143	
	A15	A151, A152				A151	A151	A151, A152		A152		A152	

Appendix B
Rates Given by Experts to Milling Machines Alternatives

© Springer International Publishing AG 2017
G. Alor-Hernández and R. Valencia-García (eds.), *Current Trends on Knowledge-Based Systems*, Intelligent Systems Reference Library 120, DOI 10.1007/978-3-319-51905-0

Appendix B: Rates Given by Experts to Milling Machines Alternatives

	E1			E2			E3			E4			E5			E6			E7			E8		
	X	Y	Z	X	Y	Z	X	Y	Z	X	Y	Z	X	Y	Z	X	Y	Z	X	Y	Z	X	Y	Z
A111	VG	VG	P	VG	G	R	VG	VG	R	VG	G	VG	VG	VG	R	G	G	VG	G	R	G	VG	VG	G
A112	G	G	P	G	R	R	VG	G	R	VG	R	VG	G	P	E	VG	R	VG	G	G	G	G	P	G
A121	G	G	P	G	G	VG	VG	VG	VG	R	R	VG	VG	VG	E	VG	VG	VG	VG	VG	VG	E	E	E
A122	R	R	P	R	R	G	VG	VG	G	VG	VG	VG	VG	VG	E	G	G	G	VG	VG	G	E	E	G
A123	G	G	P	R	R	G	VG	VG	G	G	G	G	E	E	E	G	G	G	G	G	G	VG	VG	R
A124	R	R	P	P	P	G	VG	VG	G	G	G	G	E	E	E	G	R	G	G	G	G	E	E	G
A125	M	L	L	M	M	L	M	M	M	L	M	L	M	M	L	R	L	R	L	L	L	L	L	L
A131	R	R	P	R	R	R	VG	G	R	G	VG	R	VG	VG	R	L	R	L	VG	VG	E	G	G	R
A132	R	R	P	P	P	P	VG	VG	G	G	G	R	G	G	VG	G	R	G	VG	G	E	E	E	G
A133	R	R	P	VG	G	VG	VG	VG	VG	G	VG	E	G	R	VG	R	R	VG	G	G	E	E	E	R
A134	R	R	R	VG	R	G	VG	VG	G	G	VG	E	G	G	R	G	R	VG	VG	G	VG	VG	R	VG
A135	R	R	R	VG	B	VG	VG	VG	VG	R	VG	E	VG	G	E	R	G	VG	VG	G	VG	G	P	P
A136	G	G	R	G	R	VG	VG	VG	VG	R	R	R	G	VG	VG	G	R	R	G	VG	VG	P	G	G
A137	R	R	G	R	G	VG	R	R	VG	G	G	VG	M	G	R	L	G	VG	VG	G	VG	G	G	VG
A141	VL	VL	P	VL	VL	VL	L	L	L	H	H	M	M	M	M	L	L	L	M	M	M	L	L	VL
A142	VL	VL	VL	L	L	M	L	L	M	M	H	M	M	M	L	L	M	L	L	L	L	L	L	H
A143	L	L	VL	M	M	L	L	L	L	H	H	M	M	M	M	L	M	L	M	M	M	M	M	M
A144	M	M	L	L	L	L	M	M	L	H	H	H	M	M	M	R	L	L	L	L	L	M	M	M
A151	G	G	G	R	R	G	VG	VG	G	R	R	VG	G	G	R	G	R	R	VG	VG	VG	E	E	E
A152	G	G	G	G	G	VG	G	R	VG	G	G	VG	G	G	VG	R	R	G	G	G	VG	E	E	E

References

1. Karwowski, W.: Handbook of Standards and Guidelines in Ergonomics and Human Factors (Human Factors/Ergonomics). L. Erlbaum Associates Inc. (2005)
2. Lee, J.D.: Human factors and ergonomics in automation design. In: Handbook of Human Factors and Ergonomics, 3rd edn, pp. 1570–1596 (2006)
3. Genaidy, A., et al.: Work compatibility: an integrated diagnostic tool for evaluating musculoskeletal responses to work and stress outcomes. Int. J. Ind. Ergon. **35**(12), 1109–1131 (2005)
4. Genaidy, A., Salem, S., Karwowski, W., Paez, O., Tuncel, S.: The work compatibility improvement framework: an integrated perspective of the human-at-work system. Ergonomics **50**(1), 3–25 (2007)
5. Bruseberg, A.: The design of complete systems: providing human factors guidance for COTS acquisition. Reliab. Eng. Syst. Saf. **91**(12), 1554–1565 (2006)
6. Ruckart, P.Z., Burgess, P.A.: Human error and time of occurrence in hazardous material events in mining and manufacturing. J. Hazard. Mater. **142**(3), 747–753 (2007)
7. Siemieniuch, C.E., Sinclair, M.A.: Systems integration. Appl. Ergon. **37**(1), 91–110 (2006)
8. Stanton, N.A.: Hierarchical task analysis: developments, applications, and extensions. Appl. Ergon. **37**(1), 55–79 (2006)
9. Kesseler, E., Knapen, E.G.: Towards human-centred design: two case studies. J. Syst. Softw. **79**(3), 301–313 (2006)
10. Besnard, D., Cacitti, L.: Interface changes causing accidents. An empirical study of negative transfer. Int. J. Hum. Comput. Stud. **62**(1), 105–125 (2005)
11. Vieira, E.R., Kumar, S.: Occupational risks factors identified and interventions suggested by welders and computer numeric control workers to control low back disorders in two steel companies. Int. J. Ind. Ergon. **37**(6), 553–561 (2007)
12. Balogh, I., Ohlsson, K., Hansson, G.Å., Engström, T., Skerfving, S.: Increasing the degree of automation in a production system: Consequences for the physical workload. Int. J. Ind. Ergon. **36**(4), 353–365 (2006)
13. Maldonado-Macias, A., Alvarado, A., Garcia, J.L., Balderrama, C.O.: Intuitionistic fuzzy TOPSIS for ergonomic compatibility evaluation of advanced manufacturing technology. Int. J. Adv. Manuf. Technol. **70**(9–12), 2283–2292 (2014)
14. Chan, P., Chan, F., Chan, H., Humphreys, M.: An integrated fuzzy approach for the selection of manufacturing technologies. Int. J. Adv. Manuf. Technol. **27**(7–8), 747–758 (2006)
15. Tavana, M., Khalili-Damghani, K., Abtahi, A.-R.: A hybrid fuzzy group decision support framework for advanced-technology prioritization at NASA. Expert Syst. Appl. **40**(2), 480–491 (2013)
16. Chuu, S.-J.: Group decision-making model using fuzzy multiple attributes analysis for the evaluation of advanced manufacturing technology. Fuzzy Sets Syst. **160**(5), 586–602 (2009)
17. Chuu, S.-J.: Selecting the advanced manufacturing technology using fuzzy multiple attributes group decision making with multiple fuzzy information. Comput. Ind. Eng. **57**(3), 1033–1042 (2009)
18. Ramnath, B.V., Kumar, C.S., Mohamed, G.R., Venkataraman, K., Elanchezhian, C., Sathish, S.: Analysis of occupational safety and health of workers by implementing ergonomic based kitting assembly system. Procedia Eng. **97**, 1788–1797 (2014)
19. Neumann, W.P., Medbo, L.: Ergonomic and technical aspects in the redesign of material supply systems: big boxes vs. narrow bins. Int. J. Ind. Ergon. **40**(5), 541–548 (2010)
20. Battini, B., Faccio, M., Persona, A., Sgarbossa, F., et al.: New methodological framework to improve productivity and ergonomics in assembly system design. Int. J. Ind. Ergon. (Elsevier Science) **41**(1), 30–42 (2011)

21. Jayaram, C., Jayaram, U., Shaikh, S., Kim, I., Palmer, Y.: Introducing quantitative analysis methods into virtual environments for real-time and continuous ergonomic evaluations. Comput. Ind. **57**(3), 283–296 (2006)
22. Rossi, D., Bertoloni, E., Fenaroli, M., Marciano, F., Alberti, M.: A multi-criteria ergonomic and performance methodology for evaluating alternatives in 'manuable' material handling. Int. J. Ind. Ergon. **43**(4), 314–327 (2013)
23. Battini, D., Persona, A., Sgarbossa, F.: Innovative real-time system to integrate ergonomic evaluations into warehouse design and management. Comput. Ind. Eng. **77**, 1–10 (2014)

Printed in the United States
By Bookmasters